T0215829

Lecture Notes in Computer Science 10950

Commenced Publication in 1973
Founding and Former Series Editors:
Gerhard Goos, Juris Hartmanis, and Jan van Leeuwen

FoLLI Publications on Logic, Language and Information
Subline of Lectures Notes in Computer Science

More information about this series at http://www.springer.com/series/7407

Annie Foret · Greg Kobele
Sylvain Pogodalla (Eds.)

Formal Grammar 2018

23rd International Conference, FG 2018
Sofia, Bulgaria, August 11–12, 2018
Proceedings

 Springer

Editors
Annie Foret
IRISA
Université de Rennes 1
Rennes
France

Sylvain Pogodalla
LORIA/Inria Nancy
Villers-lès-Nancy
France

Greg Kobele
Institut für Linguistik
Universität Leipzig
Leipzig, Sachsen
Germany

ISSN 0302-9743   ISSN 1611-3349   (electronic)
Lecture Notes in Computer Science
ISBN 978-3-662-57783-7   ISBN 978-3-662-57784-4   (eBook)
https://doi.org/10.1007/978-3-662-57784-4

Library of Congress Control Number: 2018947464

LNCS Sublibrary: SL1 – Theoretical Computer Science and General Issues

This Springer imprint is published by the registered company Springer-Verlag GmbH,
DE part of Springer Nature
The registered company address is: Heidelberger Platz 3, 14197 Berlin, Germany

Preface

The Formal Grammar conference series (FG) provides a forum for the presentation of new and original research on formal grammar, mathematical linguistics, and the application of formal and mathematical methods to the study of natural language. Themes of interest include, but are not limited to:

- Formal and computational phonology, morphology, syntax, semantics, and pragmatics
- Model-theoretic and proof-theoretic methods in linguistics
- Logical aspects of linguistic structure
- Constraint-based and resource-sensitive approaches to grammar
- Learnability of formal grammar
- Integration of stochastic and symbolic models of grammar
- Foundational, methodological, and architectural issues in grammar and linguistics
- Mathematical foundations of statistical approaches to linguistic analysis

Previous FG meetings were held in Barcelona (1995), Prague (1996), Aix-en-Provence (1997), Saarbrücken (1998), Utrecht (1999), Helsinki (2001), Trento (2002), Vienna (2003), Nancy (2004), Edinburgh (2005), Malaga (2006), Dublin (2007), Hamburg (2008), Bordeaux (2009), Copenhagen (2010), Ljubljana (2011), Opole (2012), Düsseldorf (2013), Tübingen (2014), Barcelona (2015), Bolzano-Bozen (2016), and Toulouse (2017).

FG 2018, the 23rd conference on Formal Grammar, was held in Sofia, Bulgaria, during August 11–12, 2018. The conference consisted in a special session, dedicated to the memory of Richard T. Oehrle, who passed away in 2018, and seven contributed papers selected from 11 submissions. The present volume includes the contributed papers.

We would like to thank the people who made the 23rd FG conference possible: the invited speakers, the members of the Program Committee, and the organizers of ESSLLI 2018, with which the conference was colocated.

August 2018

Annie Foret
Gerg Kobele
Sylvain Pogodalla

Organization

Program Committee

Jane Chandlee	Haverford College, USA
Berthold Crysmann	CNRS - LLF (UMR 7110), Paris-Diderot, France
Philippe de Groote	Inria Nancy – Grand Est, France
Nissim Francez	Technion - IIT, Israel
Thomas Graf	Stony Brook University, USA
Laura Kallmeyer	Heinrich-Heine-Universität Düsseldorf, Germany
Makoto Kanazawa	Hosei University, Japan
Stepan Kuznetsov	Steklov Mathematical Institute, Russian Federation
Robert Levine	Ohio State University, USA
Glyn Morrill	Universitat Politècnica de Catalunya, Spain
Stefan Müller	Freie Universität Berlin, Germany
Mark-Jan Nederhof	University of St Andrews, UK
Rainer Osswald	Heinrich-Heine-Universität Düsseldorf, Germany
Christian Retoré	Université de Montpellier and LIRMM-CNRS, France
Mehrnoosh Sadrzadeh	Queen Mary University of London, UK
Manfred Sailer	Goethe University Frankfurt, Germany
Edward Stabler	UCLA and Nuance Communications, USA
Jesse Tseng	CNRS, France
Oriol Valentín	Universitat Politècnica de Catalunya, France
Christian Wurm	Heinrich-Heine-Universität Düsseldorf, Germany
Ryo Yoshinaka	Tohoku University, Japan

Standing Committee

Annie Foret	IRISA, University of Rennes 1, France
Greg Kobele	Universität Leipzig, Germany
Sylvain Pogodalla	Inria Nancy – Grand Est, France

Contents

Feature Resolution by Lists:
The Case of French Coordination

Gabrielle Aguila-Multner[1]([✉]) and Berthold Crysmann[2]

[1] U Paris-Diderot, Paris, France
gabrielle.aguilamultner@gmail.com
[2] Laboratoire de linguistique formelle, CNRS & U Paris-Diderot, Paris, France
berthold.crysmann@gmail.com

Abstract. In this paper, we shall address resolution of gender and person in French coordination and suggest that a list-based encoding of feature values provides for a very simple and intuitive resolution mechanism in coordinate structures by means of simple list concatenation, while it leaves the treatment of agreement in head-compositional structures entirely unaffected. We shall discuss the implementation of this approach in the context of an emerging computational HPSG of French based on the LinGO Grammar Matrix (Bender et al. 2002), and argue that the problem at hand calls for concatenation by recursive copying (Emerson 2017), as opposed to difference lists (Clocksin and Mellish 1981). Finally, we conclude that the list-based encoding of person and gender values can act as a drop-in replacement for the standard sort-based encoding, since it is not only more flexible in the treatment of feature resolution, but also bears the further potential of representing more elaborate person systems, like the inclusive/exclusive distinction.

Keywords: Coordination · Feature resolution · Person hierarchy Gender · TDL

1 Feature Resolution in Coordination

Probably the most basic function of coordination is to combine individuals or events into aggregates. With individuals this typically creates aggregates that are treated as plurals, e.g. for the purposes of agreement. E.g. consider the examples in (1) and (2).

(1) Le chien et le chat dorment.
the dog.SG and the cat.SG sleep.PRS.3PL

'The dog and the cat sleep.'

The research reported on this paper has been partially carried out within the excellency cluster (LabEx) "Empirical Foundations in Linguistics", supported by a public grant overseen by the French National Research Agency (ANR) as part of the "Investissements d'Avenir" program (reference: ANR-10-LABX-0083).

A. Foret et al. (Eds.): FG 2018, LNCS 10950, pp. 1–15, 2018.
https://doi.org/10.1007/978-3-662-57784-4_1

(2) Le chien et le chat endormis se réveillent.
      the dog.SG and the cat.SG asleep.PL.M REFL awaken.PRS.3PL
      'The dog and the cat that were asleep are waking up.'

While the morphosyntactic number value of aggregates standardly reflects their semantic plurality (see, however, An and Abeillé 2017 for closest conjunct agreement in French NPs), there is no a priori expectation as to gender or person values of the coordinate structure unless, of course, they are composed of individuals of the same kind.

Languages like English do not show gender agreement in the plural, whereas French does, as illustrated by the agreement between subject and verb (3) or between subject and a predicative adjective (4).

(3) a. Les chevaux   sont   partis.
         the horse(M).PL be.PRS.3PL leave.PTCP.PL.M
         'The horses left.'

     b. Les tortues   sont   parties.
         the turtle(F).PL be.PRS.3PL leave.PTCP.PL.F
         'The turtles left.'

(4) a. Les frelons   sont   dangereux.
         the hornet(M).PL be.PRS.3PL dangerous.PL.M
         'Hornets are dangerous.'

     b. Les guêpes   sont   dangereuses.
         the wasp(F).PL be.PRS.3PL dangerous.PL.F
         'Wasps are dangerous.'

Gender agreement carries over to coordinate structures, as shown in (5).

(5) a. Le cheval   et l'âne   sont   partis.
         the horse(M) and the donkey(M) be.PRS.3PL leave.PTCP.PL.M
         'The horse and the donkey left.'

     b. La tortue   et la salamandre   sont   parties.
         the turtle(F) and the salamander(F) be.PRS.3PL leave.PTCP.PL.F
         'The turtle and the salamander left.'

For coordination to be functional, in a linguistic sense, there need to be resolution strategies to determine agreement not only in the case of matching gender (or person) specifications, as in (5), but also in case of mismatch.

1.1   Gender Resolution

The resolution of gender in French follows a pattern illustrated below. For a typological survey of gender systems and resolution strategies see Corbett (1991).

(6) a. Les juments   et les ânesses   sont   parties.
         the mare(F).PL and the donkey(F).PL be.PRS.3PL leave.PTCP.PL.F
         'The mares and the female donkeys left.'

     b. * Les juments   et les ânesses   sont   partis.
         the mare(F).PL and the donkey(F).PL be.PRS.3PL leave.PTCP.PL.M

(7) a. Les chevaux   et les ânesses   sont   partis.
         the horse(M).PL and the donkey(F).PL be.PRS.3PL leave.PTCP.PL.M
         'The horses and the female donkeys left.'

     b. * Les chevaux   et les ânesses   sont   parties.
         the horse(M).PL and the donkey(F).PL be.PRS.3PL leave.PTCP.PL.F

(8) a. Les juments   et les ânes   sont   partis.
         the mare(F).PL and the donkey(M).PL be.PRS.3PL leave.PTCP.PL.M
         'The mares and the donkeys left.'

     b. * Les juments   et les ânes   sont   parties.
         the mare(F).PL and the donkey(M).PL be.PRS.3PL leave.PTCP.PL.F

(9) a. Les chevaux   et les ânes   sont   partis.
         the horse(M).PL and the donkey(M).PL be.PRS.3PL leave.PTCP.PL.M
         'The horses and the donkeys left.'

     b. * Les chevaux   et les ânes   sont   parties.
         the horse(M).PL and the donkey(M).PL be.PRS.3PL leave.PTCP.PL.M

As can be seen, any occurrence of a masculine inside the coordinate structure resolves to masculine for the entire coordination, and only coordinations of exclusively feminine NPs (6) show feminine agreement. This is true at any level of embedding inside the coordinate structure, as example (10) testifies, and the constraints on agreement hold locally, as well as across non-local dependencies, as illustrated by the relative clause in (11).

(10) a. Les juments,   les ânesses   et les poneys   sont
         the mare(F).PL the donkey(F).PL and the pony(M).PL be.PRS.3PL
         partis.
         leave.PTCP.PL.M
         'The mares, the female donkeys and the ponies left.'

     b. * Les juments,   les ânesses   et les poneys   sont
         the mare(F).PL the donkey(F).PL and the pony(M).PL be.PRS.3PL
         parties.
         leave.PTCP.PL.F

(11) a. Le chien   et la tortue, qui étaient   endormis, se sont
         the dog(M) and the turtle(F) who be.IPFV.3PL asleep.M.PL be.PRS.3PL
         réveillés.
         awaken.PTCP.PL.M
         'The dog and the turtle, who were asleep, woke up.'

     b. * Le chien   et la tortue, qui étaient   endormies, se sont
         the dog(M) and the turtle(F) who be.IPFV.3PL asleep.F.PL be.PRS.3PL
         réveillées.
         awaken.PTCP.PL.F

1.2 Person Resolution

Person resolution strategies are somewhat more complex than gender resolution strategies owing to the ternary distinction of person values in French. Person agreement is illustrated for simple non-coordinated subjects in examples (12), while the resolution pattern in coordinate structures can be observed in (13–15):

(12)   a.   Nous nous entendons   bien.
            1PL   get.along.PRS.1PL well

            'We get along well.'

       b.   Vous vous entendez   bien.
            2PL   get.along.PRS.2PL well

            'You get along well.'

       c.   Elles s'entendent   bien.
            3PL.F get.along.PRS.3PL well

            'They get along well.'

(13)   a.   Toi et   moi allons   bien nous entendre.
            you and I   will.PRS.1PL well get.along.INF.1PL

            'You and I will get along well.'

       b. * Toi et   moi allez   bien vous entendre.
            you and I   will.PRS.2PL well get.along.INF.2PL

       c. * Toi et   moi vont   bien s'entendre.
            you and I   will.PRS.3PL well get.along.INF.3PL

(14)   a.   Les enfants et   moi allons   bien nous entendre.
            the child.PL and I   will.PRS.1PL well get.along.INF.1PL

            'The children and I will get along well.'

       b. * Les enfants et   moi vont   bien s'entendre.
            the child.PL and I   will.PRS.3PL well get.along.INF.3PL

(15)   a.   Toi et   les enfants allez   bien vous entendre.
            you and the child.PL will.PRS.2PL well get.along.INF.2PL

            'You and the children will get along well.'

       b. * Toi et   les enfants vont   bien s'entendre.
            you and the child.PL will.PRS.3PL well get.along.INF.3PL

The generalisation can be formulated in terms of the person hierarchy $(1 > 2 > 3)$:

– any first person conjunct triggers first person agreement;
– in the absence of any first person conjunct, any second person conjunct triggers second person agreement;
– otherwise (i.e. if all conjuncts are third person), third person agreement is used.

Once again, neither the depth of embedding in the coordinate structure (16) nor the locality of the agreement relation (17) seem to affect this pattern.

(16) a.   Les enfants, les parents   et   moi nous entendons   bien.
            the child.PL the parent.PL and I   get.along.PRS.1PL well

            'The children, the parents, and I get along well.'

      b. * Les enfants, les parents   et   moi s'entendent   bien.
            the child.PL the parent.PL and I   get.along.PRS.3PL well

(17) a.   Les enfants et   moi, qui   nous sommes rencontrés
            the child.PL and I   who be.PRS.1PL   meet.PTCP.PL.M

            hier,   nous entendons   bien.
            yesterday get along.PRS.1PL well

            'The children and I who have met yesterday get along well.'

      b. * Les enfants et   moi, qui se sont   rencontrés
            the child.PL and I   who be.PRS.3PL meet.PTCP.PL.M

            hier,   s'entendent   bien.
            yesterday get along.PRS.3PL well

The pattern for person resolution we observe for French is actually more widely attested across languages and commonly referred to in the context of the person hierarchy: e.g. English antecedent-anaphora agreement follows this pattern (Zwicky 1977), and so does subject-verb agreement in languages such as German or Russian (King and Dalrymple 2004).

1.3   Discussion

Lexical-Functional Grammar uses rather sets to represent coordinations in f-structure. Properties imposed on the set can be distributed over the members of the set, e.g. case specifications, or not, as e.g. person, number or gender specifications (Dalrymple and Kaplan 2000; King and Dalrymple 2004). Since these sets tend to be flat, membership constraints may suffice to percolate non-distributive features onto set members.[1]

HPSG does not recognise an intermediate level of representation such as f-structure but rather builds up semantics in parallel with syntactic structure. E.g. in MRS (Copestake et al. 2005), coordinations of individuals are represented as a group individual (together with its quantifier) that embeds the semantic contribution of its left and right daughters via the L-INDEX and R-INDEX features respectively (cf. (18)). The HOOK features INDEX and LTOP, which define the syntax-semantics interface (Copestake et al. 2001), however, solely expose the index and label of the coordinate structure as a whole, as illustrated by the

---

[1] Nevertheless, distribution of features in LFG will need to differentiate according to feature values, making the statement of resolution quite clumsy. E.g. feminine gender values will be distributive, whereas masculine values on the coordinate structure will only require membership on one of the f-structure sets contributed by the conjunct-daughters. One can imagine that such a regime will become even more unwieldy, once we move to a tri-fold resolution scheme, as observed with person.

sample MRS in (18). As a masculine noun triggering masculine agreement of the entire coordination can be embedded arbitrarily deep in a coordination of feminines (see example (10)), access to any person or gender features of conjuncts would necessitate traversing the MRS graph, e.g. by means of functional uncertainty, a solution that runs counter the idea of a lean interface between syntax and semantics, as advanced by Copestake et al. (2001).

(18)   MRS for *la girafe et l'éléphant* (quantifiers omitted)

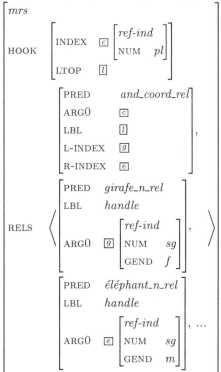

A straightforward alternative solution would shift the burden to the syntax: for gender alone, a rather brute-force approach would expand what is now a single coordination rule into three, projecting *fem* to the coordination in the case of all feminine daughters, projecting *mas* from a masculine left daughter, and finally, projecting *mas* from the right daughter with a feminine left daughter[2]. The same needs of course to be done for person, yet with combinations of three values to be taken care of, instead of just two. Worse, since both gender and person may

---

[2] This last restriction avoids spurious ambiguities with all masculine coordinations, using three rules for the four logical combinations.

need to be resolved at the same time, as illustrated by the example in (19), the rules may actually multiply out, in the worst case.[3]

(19)   Elle,   son frère   et   moi   nous sommes bien entendus.
       3SG.F   her brother(M.SG)   and 1.SG be.PRS.1PL   well get.along.PTCP.PL.M
       'She, her brother, and I got on well.'

This is not just uneconomical, but the need to multiply out rules for combinations of feature values is something that unification grammar set out to eliminate in order to improve over CFGs. Furthermore, enumeration of combinations will end up obfuscating the linguistically rather clear person and gender hierarchies that govern resolution.

A rather radical approach to feature resolution has been proposed by Sag (2003): instead of unification, he proposed subsumption checks, an approach that bears some similarity to Ingria (1990). This change in the underlying logic of typed feature formalism, however, has not been widely adopted. Furthermore, there is no implementation to date that supports this. Fortunately, as we shall show below, a simple extension to the representation of gender and person features is sufficient to address the issue of feature resolution using a standard unification formalism.

In the following section, we shall therefore develop a theory of feature resolution that crucially distinguishes between a feature itself and its resolution potential. More concretely, we shall enrich the representation of *per* and *gend* values in order to distinguish between e.g. *being* first person or *being* masculine and containing a first person or masculine. We shall show that once the signature of these features is slightly enriched, feature resolution in coordinate structures can be done deterministically. This move leaves untouched standard phrase structure rules targeting entire INDEX values, including all of person, number, and gender features, whereas coordinate structures will have the required flexibility to determine the resolution potential for each feature either holistically, as in the case of semantically motivated number (*pl*), or else in terms of a syntactic resolution strategy.

---

[3] These observations are only true, in a strict sense, for pure unification formalisms. In systems like Trale (Penn 2004) the disjunctions between rules could be relegated to attached relational constraints or even better, implicational constraints, as pointed out to us by an anonymous reviewer. However, once a general solution has been found for formalisms without these more elaborate constraints, it certainly helps towards closing the gap in expressiveness between the two competing approaches to HPSG implementation.

2   Analysis

2.1   The Basic Approach: Using Lists to Express Existential Constraints

To apply this idea to gender resolution, we first enrich the type *gend* with a feature M taking a list as its value, cf. (20). The type *mas* is then constrained to have a non-empty M list, while the type *fem* is required to have an empty M list

(20)   Type constraints for gender

$$
\begin{bmatrix} gend \\ \text{M} \quad list \end{bmatrix}
$$

$$
\begin{bmatrix} mas \\ \text{M} \quad \langle [\,], ... \rangle \end{bmatrix} \begin{bmatrix} fem \\ \text{M} \quad \langle \, \rangle \end{bmatrix}
$$

Similarly, we enrich the type *per* with the two list-valued features ME and YOU, cf. (21); the type *1st* requires a non-empty ME list, but does not constrain the YOU list; the type *2nd* requires a non-empty YOU list and an empty ME list; finally the type *3rd* requires both lists to be empty.

(21)   Type constraints for person

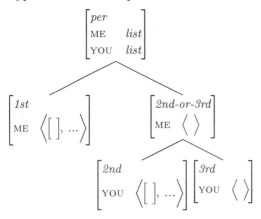

While this elaborate structure will change nothing with respect to standard projection of INDEX values in head-compositional structures, we gain added flexibility when dealing with coordination: recall that coordinations of individuals (or events) introduce their own INDEX variable, which represents the aggregate. Thus, what needs to be done to capture feature resolution is to determine the PER and GEND values of the group variable on the basis of the respective features of the group members, whereas the NUM value transparently represents plural semantics of the aggregate.

Given our list representation, we expand the coordination rule types of the Grammar Matrix[4] (Drellishak and Bender 2005) by the constraint in (22) above, enabling us to directly compute the values of agreement features by list concatenation.

(22)

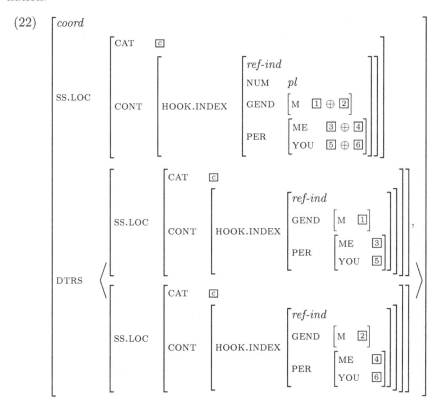

Furthermore, to ensure lists are lexically well-terminated, we constrain lexical types accordingly: (23) defines *basic-noun-lex*, a type from which lexical types fro nouns and pronouns inherit.[5]

---

[4] The LinGO Grammar Matrix (Bender et al. 2002) is a starter kit for the development of implemented HPSG-style grammars, which has been distilled, originally, from the implemented grammars of English (Copestake and Flickinger 2000) and Japanese (Siegel and Bender 2002). On the syntactic side, the Matrix provides type definitions for grammars developed in the spirit of Ginzburg and Sag (2000). With respect to semantics, the Matrix provides compositional principles for semantics construction in Minimal Recursion Semantics (Copestake et al. 2005), ensuring both reversibility and cross-linguistic interoperability.

[5] The list type *0-1-list* denotes a list with length of at most 1. It is straightforwardly defined in TDL as follows:

i. *0-1-list* := *list*.

ii. *1-nelist* := *0-1-list* ∧ *nelist* ∧ [REST *elist*].

iii. *elist* := *0-1-list*.

(23) *basic-noun-lex* →

$$
\left[\text{SS.LOC}\begin{bmatrix}\text{CAT}&\begin{bmatrix}\text{HD}&\textit{noun}\end{bmatrix}\\\text{CONT}&\begin{bmatrix}\text{HOOK}&\begin{bmatrix}\text{INDEX}&\begin{bmatrix}\text{GEND}&\begin{bmatrix}\text{M}&\textit{0-1-list}\end{bmatrix}\\\text{PER}&\begin{bmatrix}\text{ME}&\textit{0-1-list}\\\text{YOU}&\textit{0-1-list}\end{bmatrix}\end{bmatrix}\end{bmatrix}\end{bmatrix}\end{bmatrix}\right]
$$

With these basic constraints in place, we will obtain the following results for gender in a coordination of two NPs:

(24) a.
$$\left[\text{GEND}\quad m\left[\text{M}\left\langle[\,]\right\rangle\right]\right]+\left[\text{GEND}\quad m\left[\text{M}\left\langle[\,]\right\rangle\right]\right]$$
$$=\left[\text{GEND}\quad\left[\text{M}\left\langle[\,],[\,]\right\rangle\right]\right]$$

b.
$$\left[\text{GEND}\quad m\left[\text{M}\left\langle[\,]\right\rangle\right]\right]+\left[\text{GEND}\quad f\left[\text{M}\left\langle\,\right\rangle\right]\right]$$
$$=\left[\text{GEND}\quad\left[\text{M}\left\langle[\,]\right\rangle\right]\right]$$

c.
$$\left[\text{GEND}\quad f\left[\text{M}\left\langle\,\right\rangle\right]\right]+\left[\text{GEND}\quad m\left[\text{M}\left\langle[\,]\right\rangle\right]\right]$$
$$=\left[\text{GEND}\quad\left[\text{M}\left\langle[\,]\right\rangle\right]\right]$$

d.
$$\left[\text{GEND}\quad f\left[\text{M}\left\langle\,\right\rangle\right]\right]+\left[\text{GEND}\quad f\left[\text{M}\left\langle\,\right\rangle\right]\right]$$
$$=\left[\text{GEND}\quad\left[\text{M}\left\langle\,\right\rangle\right]\right]$$

Note that the result of list concatenation is underspecified as to the *gend* type[6], but the resulting non-empty M lists in (24a-c) are only compatible with the type constraints of [GEND *mas*], thus triggering masculine agreement, whereas the empty list in (24d) is only compatible with the type constraint for [GEND *fem*].

---

[6] The LKB, unlike Trale, does not allow inference from features to the types that introduce or constrain them.

2.2   A Closer Look at List Concatenation in TDL

Pure unification formalisms[7], such as the LinGO LKB (=Linguistic Knowledge Builder; Copestake 2002) do not recognise lists as primitive data structures, nor do they provide any specific list operations like, e.g. member/2 or append/3. Rather, lists are encoded as feature structures in FIRST/REST (or HD/TL) notation, which provides easy access for push and pop operations. List concatenation, however, is typically done via difference lists (Clocksin and Mellish 1981), which maintain an additional pointer to the end of an open list onto which additional lists can be unified.

In a first attempt, we have therefore used difference lists to concatenate the respective person and gender lists, such as GEND.M or PER.ME in coordinate structures. The problem that soon transpired was that difference lists are open lists by necessity, such that any attempt to constrain to a non-empty difference list of indeterminate length (25a) was even successful with empty difference lists (25b), unifying a list element onto the pointer to the end of the list, as shown in (25c).[8]

(25)   a. Non-empty difference list underspecified for length

$$
\begin{bmatrix}
\text{LIST} & \begin{bmatrix} \text{FIRST} & [\ ] \\ \text{REST} & \mathit{list} \end{bmatrix} \\
\text{LAST} & \mathit{list}
\end{bmatrix}
$$

   b. Empty difference list

$$
\begin{bmatrix}
\text{LIST} & \boxed{l} \\
\text{LAST} & \boxed{l}
\end{bmatrix}
$$

   c. Unifying a non-empty list onto an empty difference list

$$
\begin{bmatrix}
\text{LIST} & \boxed{l}\begin{bmatrix} \text{FIRST} & [\ ] \\ \text{REST} & \mathit{list} \end{bmatrix} \\
\text{LAST} & \boxed{l}
\end{bmatrix}
$$

The only possible solution would have been to terminate this pointer at some point, which proved hard to do in a general and principled fashion. Furthermore, it was difficult to express length constraints on difference lists, as used, e.g. in (23) above.

---

[7] TDL (=Type Description Language; Krieger 1996.) was the original description language of the PAGE system (Uszkoreit et al. 1994) and currently is the standard description language for typed feature structure grammar development and runtime platforms within the DELPH-IN collaboration, such as the LKB (Copestake 2002), Pet (Callmeier 2000), and Ace (Crysmann and Packard 2012). Grammars specified in TDL include the English Resource Grammar (Copestake and Flickinger 2000) among several others, as well as the LinGO Grammar Matrix (Bender et al. 2002).

[8] Note that checking for cyclic feature structures – a check which the LKB indeed performs—will not provide a solution: once we need to underspecify the length of the list, reentrancy between the REST and LAST cannot be stated.

Fortunately, Emerson (2017) has recently proposed a method to perform concatenation directly on lists, using recursive copying of list members. Following his proposal, we implemented the constraint in (22) using the list definitions in (26), yielding an implementation of the coordinate structure constraint as in (27).

(26)   List concatenation in TDL (Emerson 2017)

   a. *list-copy* := *list* $\wedge$

$$\begin{bmatrix} \text{COPY} & \textit{list} \\ \text{NEXT} & \textit{list} \end{bmatrix}$$

   b. *nelist-copy* := *list-copy* $\wedge$ *nelist* $\wedge$

$$\begin{bmatrix} \text{FIRST} & \boxed{f} \\ \text{REST} & \begin{bmatrix} \textit{list-copy} \\ \text{COPY} & \boxed{r} \\ \text{NEXT} & \boxed{n} \end{bmatrix} \\ \text{COPY} & \begin{bmatrix} \text{FIRST} & \boxed{f} \\ \text{REST} & \boxed{r} \end{bmatrix} \\ \text{NEXT} & \boxed{n} \end{bmatrix}$$

   c. *elist-copy* := *list-copy* $\wedge$ *elist* $\wedge$

$$\begin{bmatrix} \text{COPY} & \boxed{n} \\ \text{NEXT} & \boxed{n} \end{bmatrix}$$

The declarations in (26) faithfully replicate the proposal by Emerson (2017) for purely list-based append: the core idea is to augment a typed feature structure list representation with features for a successor list (NEXT) and a result list COPY. In essence, the FIRST/REST part of the enriched structure represents the first list, the NEXT feature the second list and the COPY holds the resulting concatenation. The type *list-copy* merely introduces the appropriate features (26a). The second clause (26b) recurses over the first list, token-identifying element by element the members of the first list with the members of COPY, the result list. Once the end of the first list has been reached, a subtype of *elist*, (26c) identifies the second list (NEXT) with the result list (COPY). The COPY feature of the entire list will thus consist of the second list, plus the elements of FIRST prepended to it member by member.

In the implementation of feature resolution in French, we consequently use Emerson-style list concatenation, as illustrated in (27). Using gender as an example, the NEXT feature of the GEND.M list of the left conjunct is equated, in coordinate structures, with the GEND.M list of the right conjunct, and the resulting list concatenation in the left daughter's GEND.M.COPY will be token-identical to the GEND.M list on the mother of the coordinating construction.

(27)
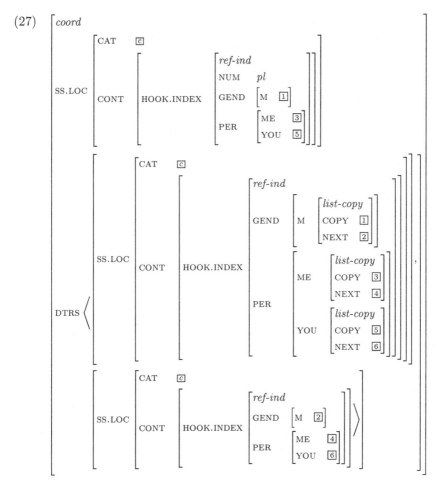

Fully in parallel to gender, the constraint in (27) equally describes concatenation of person values, broken down to ME and YOU features.

3   Conclusion

In this paper we have discussed resolution of gender and person features in French coordination and suggested to augment the representation of their values with a list-based encoding, and we have shown how this simple extension in the type signature enables us to address the issue of resolution in terms of simple list concatenation. Both the simplicity of the approach and the cross-linguistic recurrence of the phenomenon will make this solution easily applicable to a wider range of languages, both theoretically and within the context of multilingual grammar engineering.

The redundancy between type-based encoding of person and gender features and the list-based one raises the obvious question whether the latter can fully substitute for the former. As for French, we can answer this question in the affirmative, since the lists we propose clearly cover the full inventory of distinctions, yet provide the additional option of distinguishing between exclusively having some property (closed list) and containing some property (open list). Furthermore, feature resolution by concatenation constitutes a simple and uniform mode of composition. The decomposition of person into ME and YOU features bears the further potential to provide an encoding of inclusive and exclusive person distinctions in the plural, as suggested, e.g. by Anderson (1992). Finally, the present approach clearly shows that feature resolution can be done with unification alone, obviating the need for subsumption checks.

References

An, A., Abeillé, A.: Agreement and interpretation of binominals in French. In: Stefan, M. (ed.) Proceedings of the 24th International Conference on Head-Driven Phrase Structure Grammar, University of Kentucky, Lexington, pp. 26–43. CSLI Publications, Stanford (2017). http://cslipublications.stanford.edu/HPSG/2017/hpsg2017-an-abeille.pdf

Anderson, S.R.: A-Morphous Morphology. Cambridge University Press, Cambridge (1992)

Bender, E.M., Flickinger, D., Oepen, S.: The grammar matrix: an open-source starter-kit for the rapid development of cross-linguistically consistent broad-coverage precision grammar. In: John, C., Nelleke, O., Richard, S. (eds.) Proceedings of the Workshop on Grammar Engineering and Evaluation at the 19th International Conference on Computational Linguistics, pp. 8–14 (2002)

Callmeier, U.: PET—a platform for experimentation with efficient HPSG processing techniques. J. Nat. Lang. Eng. **6**(1), 99–108 (2000)

Clocksin, W.F., Mellish, C.S.: Programming in Prolog. Springer, Heidelberg (1981)

Copestake, A.: Implementing Typed Feature Structure Grammars. CSLI Publications, Stanford (2002)

Copestake, A., Flickinger, D.: An open-source grammar development environment and broad-coverage English grammar using HPSG. In: Proceedings of the Second conference on Language Resources and Evaluation (LREC-2000), Athens (2000)

Copestake, A., Flickinger, D., Pollard, C., Sag, I.: Minimal recursion semantics: an introduction. Res. Lang. Comput. **3**(4), 281–332 (2005)

Copestake, A., Lascarides, A., Flickinger, D.: An algebra for semantic construction in constraint-based grammars. In: Proceedings of the 39th Annual Meeting of the Association for Computational Linguistics, Toulouse, France (2001)

Corbett, G.G.: Gender. Cambridge University Press, Cambridge (1991)

Crysmann, B., Packard, W.: Towards efficient HPSG generation for German, a non-configurational language. In: Proceedings of the 24th International Conference on Computational Linguistics (COLING 2012), Mumbai, India, pp. 695–710 (2012)

Dalrymple, M., Kaplan, R.: Feature indeterminacy and feature resolution. Language **76**(4), 759–798 (2000)

Drellishak, S., Bender, E.M.: A coordination module for a crosslinguistic grammar resource. In: Stefan, M. (ed.) Proceedings of the 12th International Conference on Head-Driven Phrase Structure Grammar, Department of Informatics, University of Lisbon, pp. 108–128 (2005). http://cslipublications.stanford.edu/HPSG/2005/drellishak-bender.pdf

Emerson, G.: (Diff-)list appends in TDL. In: 13th DELPH-IN Summit Oslo (2017)

Ginzburg, J., Sag, I.A.: Interrogative Investigations. The Form, Meaning, and Use of English Interrogatives. CSLI Publications, Stanford (2000)

Ingria, R.J.P.: The limits of unification. In: Proceedings of the Twentyeigth Annual Meeting of the Association for Computational Linguistics, pp. 194–204 (1990)

King, T.H., Dalrymple, M.: Determiner agreement and noun conjunction. J. Linguist. **40**, 69–104 (2004)

Krieger, H.-U.: TDL—a type description language for constraint-based grammars. Saarbrücken Dissertations in Computational Linguistics and Language Technology, vol. 2. DFKI GmbH, Saarbrücken (1996)

Penn, G.: Balancing clarity and efficiency in typed feature logic through delaying. In: Proceedings of the 42nd Meeting of the Association for Computational Linguistics (ACL 2004), Barcelona, Spain, pp. 239–246 (2004). https://doi.org/10.3115/1218955.1218986. http://www.aclweb.org/anthology/P04-1031

Sag, I.A.: Coordination and underspecification. In: Kim, J.-B., Wechsler, S. (eds.) Proceedings of the 9th International Conference on Head-Driven Phrase Structure Grammar, Kyung Hee University, Seoul, 5–7 August 2002, pp. 267–291. CSLI Publications, Stanford (2003)

Siegel, M., Bender, E.M.: Efficient deep processing of Japanese. In: Proceedings of the 3rd Workshop on Asian Language Resources and International Standardization, Coling 2002 Post-conference Workshop, Taipei (2002)

Uszkoreit, H., Backofen, R., Busemann, S., Diagne, A.K., Hinkelman, E., Kasper, W., Kiefer, B., Krieger, H.-Ul., Netter, K., Neumann, G., Oepen, S., Spackman, S.P.: DISCO - an HPSG-based NLP system and its application for appointment scheduling. In: Proceedings of the 15th International Conference on Computational Linguistics (COLING 1994), Kyoto, Japan, 5–9 August, vol. 1, pp. 436–440 (1994)

Zwicky, A.: Hierarchies of person. In: Chicago Linguistic Society, vol. 13, pp. 714–733 (1977)

Paracompositionality, MWEs and Argument Substitution

Cem Bozşahin[(✉)] and Arzu Burcu Güven

Cognitive Science Department, Informatics Institute,
Middle East Technical University (ODTÜ), Ankara, Turkey
`bozsahin@metu.edu.tr`, `arzuburcuguven@gmail.com`

Abstract. Multi-word expressions, verb-particle constructions, idiomatically combining phrases, and phrasal idioms have something in common: not all of their elements contribute to the argument structure of the predicate implicated by the expression.

Radically lexicalized theories of grammar that avoid string-, term-, logical form-, and tree-writing, and categorial grammars that avoid wrap operation, make predictions about the categories involved in verb-particles and phrasal idioms. They may require singleton types, which can only substitute for one value, not just for one kind of value. These types are asymmetric: they can be arguments only. They also narrowly constrain the kind of semantic value that can correspond to such syntactic categories. Idiomatically combining phrases do not subcategorize for singleton types, and they exploit another locally computable and compositional property of a correspondence, that every syntactic expression can project its head word. Such MWEs can be seen as empirically realized categorial possibilities rather than lacuna in a theory of lexicalizable syntactic categories.

Keywords: Syntax · Semantics · CCG · Multi-word expression
Idiom · Verb-particle · Lexical insertion · Type theory

1   Introduction

A type is a set of values. When we write a syntactic type, say NP, we mean a set of expressions (values) which can substitute for that type. This type serves to distinguish some expressions from for example the set of expressions that can substitute for a VP type.

The distinction is crucial for solving the correspondence problem in syntax-semantics. For this purpose we talk about semantic types, for example e for things and t for propositions. The concepts that can substitute for semantic types are not expressions in the sense that syntactic expressions are, because

We thank Tzu-Ching Kao, Umut Özge and Mark Steedman for discussion related to the paper, and three anonymous reviewers of *FG* for commentary and sources. All misunderstanding is ours.

they are not observable, but they leverage a theory to hypothesize about the kind of semantics that these types stand for.

These two species of types are then put in a correspondence in a theory of syntax-semantics connection. The understanding is that if one substitutes a certain expression for a syntactic type, then its corresponding semantic type substitutes for a certain kind of semantic value. We know less about the semantic values; but, at the level of the correspondence problem, this is not very critical. It is however crucial to make the distinctions and propagate them in a parsing mechanism rather than solving all type-interpretation problems in one go.

We need a theory which provides explicit vocabulary and mechanism for the correspondence, to be more specific about the equal relevance of substitution for subexpressions which purportedly do not contribute to the meaning of the expression.

In the categorial grammar parlance, for which we will use Combinatory Categorial Grammar [29,30], hereafter CCG, we can exemplify the correspondence as follows, where we use the "result-first argument-next" notation:

(1)   a.   hits := $(S \backslash NP_{3s})/NP$: $\lambda x \lambda y.hit' xy$

 b.   hit := VP_{inf}/NP: $\lambda x \lambda y.hit' xy$

Some syntactic types are further narrowed down by features, such as NP_{3s} above for third person singular NP, which are, in CCG, not re-entrant.

We argue in the paper that in a radically lexicalized theory which adheres to transparency of derivations by type substitution (rather than lexical insertion), such as CCG, there are built-in degrees of freedom to support Multi-word Expressions (MWEs) and idioms without complicating the mechanism.

Paracompositionality is key to projection of their properties in a derivation. It is the idea that, in addition to the compositionality of the lexical correspondence, which is compositional partly because it relies on non-vacuous abstractions, type substitution by (i) what we call singleton types and (ii) what is called head-dependencies in the NLP literature is also compositional because it spells non-vacuous abstraction as part of the correspondence, but as something related to the contingency of the predicate, rather than the argument structure of the predicate. In a radically lexicalized grammar both sources are available in a lexical item. These types are paracompositional also in the sense that whether we have an idiom reading or compositional one is already decided by the category of the head in the derivational process.

The term *contingency* is used here in the sense of Moens and Steedman [23] where it relates to extension of happenings. In the case of events (culminations, points, processes and culminating processes), which have definite extension, it is an event modality of space, time and manner; and, in the case of states where extension is indefinite (e.g. *understand*) it is some property of the state. From now on when we use the term 'contingency' we mean something related to extension of the predicate, rather than who does what to whom in the predicate.

MWEs are expressions involving more than one word in which the properties of the expression are not determined by the composition of the properties of the constituent words, which would be the case for phrases. There is a tendency to treat them as single lexical units [10,33]; but, as we shall see, CCG does not require the single unit to be the phonological representation to the left of ':=' in the format of (1). This property of CCG naturally extends to coverage of verb-particle constructions e.g. *look the word up* as discontiguous MWEs headed by a lexical item.

Phrasal idioms and idiomatically combining phrases are classes identified by Nunberg, Sag and Wasow [24] to account for systematic variation in syntactic productivity of idioms. Typewise they will relate to singleton types (phrasal idioms) and head-word subcategorization (idiomatically combining phrases) in our formulation.

As a preview of the article, we can think of the meaning distinctions as ranging from "beans" i.e. the nounphrase *beans* itself as a category (this is what we call the singleton type); to NP_{beans} as the category of an NP headed by the word *beans*, which has wider range of substitution; and, to the polyvalent NP with the widest substitution for that type. This much is categorial grammar with type substitution. CCG as an empirical theory adds to this the claim that there is an asymmetry in the range of substitutions: the singleton types can be arguments only, and arguments of arguments and results, but never the result. We shall see that this has implications for the linguist's choice of handling syntactic productivity in a grammar.

Some implications follow: Because of paracompositionality, all expressions requiring a singleton type would involve the semantic type of a predicate, and all idiomatically combining phrases requiring a different interpretation than the compositional one would have the same consequence independent of their syntactic productivity. In short, every idiom must contain a predicate (but not necessarily a verb). We cover these implications in the article.

2 Substitution in a Derivation

In (1a), the '$/NP$' can be substituted for by certain kinds of expressions, for example *John, me, the ball, a stone in the corner*, etc. Its corresponding semantic counterpart in the logical form (LF), written after the colon, has the placeholder x which can be typed as e, to be suitably substituted for by a semantic value described above. The '$\backslash NP_{3s}$' can be substituted by narrower expressions, for example eliminating *I, you*. Because this is an indirect correspondence, its semantic counterpart y can have the same type e.

The tacit assumption of indirectness is sometimes made explicit, for example in Bach's [2] rule-by-rule hypothesis: The derivational process operates with syntactic types only, and when it applies the semantics of the rule, its semantics works only with LF objects. Quoting from Bach: "Neither type of rule has access to the representations of the other type except at the point where a translation rule corresponding to a given syntactic rule is applied." The "syntactic rule" in a lexicalized grammar such as CCG is the combinatory syntactic type of a lexical correspondence. The "translation rule" is the lexically-specified logical form, LF, as in (1).

The derivational process reveals partially derived types, for example $S\backslash NP_{3s}$: $\lambda y.hit's'y$ for (1a), if function application substitutes say *a stone* for '$/NP$', with some semantic value s'. The semantic type of such derived categories is concomitantly functional, e.g. $e \mapsto t$ for this syntactic type. *John hits* is $e \mapsto t$ too, with category $S/NP : \lambda x.hit'x\,john'$.

We can see the relevance of derived types to substitutability in a closer look at (1b). If function application substitutes for the '$/NP$' in the example, the derived category would be $VP_{\mathrm{inf}}: \lambda y.hit's'y$ in this case. This is also an $e \mapsto t$ type semantically. However, its syntax is narrower so that we can account for the expressions in (2).[1]

(2)   John persuaded Mary to/* hit/*hits the target.

The derivational process works as below, with $VP_{\text{to-inf}}$ distinct from VP_{inf}.

(3)

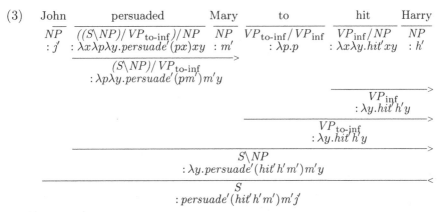

Here function application is shown in forward form ($>$) and backward form ($<$). Derivation proceeds from top to bottom in display, as standard in CCG; i.e., bottom-up as far as parsing is concerned, and one at a time. For brevity alternative derivations using function composition are not shown; their implications for constituency are discussed in Steedman references. We also eschew the slash modalities of Baldridge and Kruijff [3] to avoid digression, which can further restrict the combination possibilities of syntactic types. They are mentioned later when they are relevant to discussion. The LF contains a structured

---

[1] This is equivalent to saying that in CCG the type VP is not always an abbrevation for $S\backslash NP$, which might be the case in other brands of categorial grammars. The English facts above could be taken care of by featural distinctions such as $S_{\mathrm{inf}}, S_{\text{to-inf}}, S_{\mathrm{fin}}$ in $S\backslash NP$, rather than also positing a VP. But in ergative languages the '$\backslash NP$' does not always coincide with the same LF role as it does in English, such as in Dyirbal's control construction, where the controlled absolutive argument can be the patient NP of the transitive clause or syntactic subject of an intransitive clause, but not the ergative NP of the transitive clause. It seems to require $VP: \lambda x.pred'x$ where x's role in the controlled clause $pred'$ is determined by verbal morphology of the controlled clause; see [22] for the phenomenon. Assuming a VP cross-linguistically makes narrower predictions about control. We handle this problem elsewhere.

form, viz. the predicate-argument structure, which is written in linear notation for simplicity; for example $hit'xy$ is same as $((hit'x)y)$; i.e., it is left-associative.

In preparation for final discussion of substitution (§6) in relation to the wrapping operation, we can redraw this derivation by showing the substituting expressions as we proceed, which we do in Fig. 1.

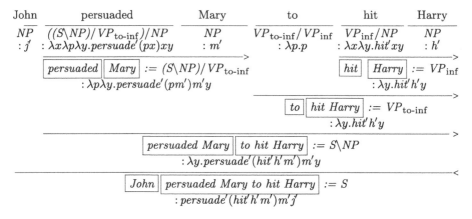

Fig. 1. Substitution of syntactic expressions for syntactic types. Boxes show segments combined. We display some one-at-a-time derivations on the same line to save space.

MWEs present a challenge for substitution in such correspondences. In Schuler and Joshi's [28]:25 words: "In the *pick .. up* example, there is no coherent meaning for *Up* such that $[\![pick\ X\ Up]\!] = Pick([\![X]\!], Up)$." They go on to show how tree-write in the form of TAG transformations, rather than string-rewrite of CFG transformations such as [27], can deliver different meanings of such expressions *after* a fully compositional tree is established for 'pick', '..' and 'up'.

In such systems, post-processing and reanalysis of a categorial surface derivation are possible, both for TAG and HPSG,[2] therefore these transformations are possible, indeed useful, to simplify large-scale grammar development.

For radically lexicalized grammars such as CCG where such options are not available, three paths to maintaining compositionality in the presence of "non-compositional" and/or idiomatic parts seem to be available:

---

[2] TAG transformations take a phrase structure tree and decompose it to elementary structures to deliver an LF. [21] is a different TAG way to incorporate meaning postulates of [25]. HPSG uses phrasal post-classification to the same effect. For example [4,27] perform it at the final stage of parsing as a semantic check on bags of predicates for idiom entries, and [17] use semantic frame identification, viz. compositional vs idiomatic, which are built in to theory. The diversity of approaches in the volume for idioms [14] is testimony to the practice that the idioms are decisive factors in polishing our theories linguistically, psychologically and computationally.

(4) a. letting the logical form change the compositional meaning,

b. introducing surface wrap,

c. reassessing the substitutability of *argument* types, to the extent that (i) they can be narrowed by head-dependencies, and (ii) the semantic contribution of some parts of the correspondence to the predicate-argument structure can be ignored in a principled way, and locally.

The problem is exacerbated by phrasal idioms which seem to have partially active syntax in some non-compositional parts, for example *kick the (proverbial/old) bucket*, but note ♯*the bucket that John kicked*, ♯*kick the great bucket in the sky*, and **the breeze was shot*. (♯ is used to indicate unavailability of idiomatic reading. The last two examples and judgments are from [27].) However, there are also phrasal idioms which are syntactically quite active, e.g. *the beans that John spilled*, and *spilling the musical/artistic/juicy beans*.

Option (4a) does not always necessitate post-processing of MWEs in CCG, but, as we shall see later in (23), it does not guarantee locality of derivations either. One way to realize it is the following:

(5) kicked := *(S\NP)/NP* : $\lambda x \lambda y.$if $head(x) = bucket'$then $die'y$ else $kick'xy$

This approach to phrasal idioms which is similar to meaning postulates for the same task such as [25] would then have to make sure that the head meaning *bucket'* has some predefined cluster of modifiers such as *proverbial* or *old*, but not much else, for example ♯*kick the bucket that overflowed*. It would also have to overextend itself to avoid the idiomatic reading in ♯*the bucket that you kicked*.

As an alternative, the type NP_{bucket} below is inspired by trainable stochastic CFGs which can distinguish argument PPs from adjunct PPs by encoding head dependencies for CFG rules, for example $VP_{\text{put}} \rightarrow V_{\text{put}}\ NP\ PP_{\text{on}}$: (We shall fix the unaccounted vacuous abstraction in it later in the paper.)

(6) kicked := *(S\NP)/NP*$_{\text{bucket}}$: $\lambda x \lambda y.die'y$

It might appear to be LF-motivated just like (5) above; but, it is actually a case of (4c/i). NP_{bucket}, meaning NP headed by *bucket*, can be made distinct from NP_{buckets} because different surface expressions can be substituted for them. (6) overgenerates for the examples given above, but it might be the right degree of freedom to exploit in the syntax-semantics correspondence of idiomatically combining MWEs such as NP_{beans} for *spill the beans*.

In the remainder of the paper, we show that option (4c/ii) has been implicit in CCG theory all along but never used, in the form of syntactic types for which only one value can substitute (Sect. 3). We call them singleton types. This way of lexical categorization and subcategorization predicts very limited syntax, but not as metalinguistic marking that [27] proposed for *kick the proverbial/old bucket*. It is due to having to enumerate different senses and contingencies of phrasal idioms (e.g. proverb bucket for senses above, also covering e.g. *when I face the proverbial bucket*), and *pick up* for MWEs. In Sect. 4 we show that idiomatically combining phrases have principled distinctions from singleton types. Head-word subcategorization such as (6) is the more promising option for them, which radically lexicalized grammars can handle without extension. There are also idioms which require analysis combining both options such as those with semantic reflexives where the referent is not part of the idiom, e.g. *I twiddle my/*his thumbs.* Section 5 covers these cases.

These findings reveal some aspects of type substitution and its projection when the expressions are not fully compositional at the level of the predicate-argument structure. As such they may have implications beyond CCG.

Finally we show that adopting option (4b) to analyze for example *pick · · · up* as *pick up* $(\cdots)_{\text{wrap}}$ overgenerates in the combinatory version of wrap (Sect. 6), and complicates the grammar with a domino-effect in the surface version of wrap; therefore, it would do more damage than good if adopted for (discontiguous) MWEs and phrasal idioms. CCG can continue to avoid all forms of wrap in the presence of all kinds of MWEs and phrasal idioms.

3   Singleton Types

A brief preview of the proposal for (4c/ii) is as follows. A singleton syntactic type self-represents because it can substitute for one value only. We designate such types with strings, such as *"up"* or *"the bucket"*; for example:

(7)   a. picked := $(S\backslash NP)/$ *"up"*$/NP$: $\lambda y \lambda x \lambda z.cause'(init'(hold_x'yz))z$

     b. kicked := $(S\backslash NP)/$ *"the bucket"* : $\lambda x \lambda y.die_x'y$

(*Init'* is a function that yields a culminating state in the sense of [23].)

We call categories in (7) 'paracompositional' to highlight the fact that, although their LF correspondence is intact so that the derivational process is transparent, they might have seemingly vacuous abstraction from the perspective of the predicate-argument structure, symbolized by the placeholders x above.[3]

However, one can make a case that this abstraction, corresponding respectively to singleton categories *"up"* and *"the bucket"*, might have a role inside the LF constants shown in primes, as contingencies. We write them for example as $die_x'y$ (as ceremonial death, reported death, etc.), rather than $die'y$. These LF 'constants' are convenient generalizations in CCG standing in for a plethora of features anyway, so it seems natural to think of them as having their own

---

[3] van der Linden [32], which is another categorial approach to idioms, allows vacuous abstractions, i.e. define semantics without mention of x in the LF of (7b). Apart from our empirical claim that they have a place in LF because they relate to contingency, vacuous abstractions seem to open ways to resource insensitivity which is unheard of in natural language; for example, the **K** combinator with its vacuous abstraction $\lambda x \lambda y.x$ can delete things from LF. We have yet to find a word or morpheme that does this; see [5]:81 for some speculation.

    [32]'s treatment of phrasal idioms such as *kick the bucket* assumes partial involvement of the head verb *kick* for the semantics of the idiom, whereas in our conception it is fully responsible for the idiom with the aid of singleton types.

abstraction. (The semantic types corresponding to these contingencies are then $\alpha \mapsto t$ for some α.)

It will be seen in Sect. 3.2 that the examples in (7) differ in their sense from *picked up the book* and *kicked the blue bucket*, therefore a separate grammar entry is empirically justified. The sense distinction is reflected explicitly in the LF, as we shall see later. Both possibilities for substitution, for the syntactic type and for its placeholder in the LF, are principally restricted by CCG.

Singletons also engender a way for such entries to be morphologically more transparent, for example by being susceptible to inflection, e.g. *picking*, by providing a segmental alternative to contiguous but MWE *pick up* $\cdots$, which would need a morphological pointer for inflection, as noted by [27,33] for their analyses. Nunberg, Sag and Wasow's [24] dichotomy between phrasal idioms and idiomatically combining items also vanishes, because of the singleton types and head-word subcategorized argument types. The distinction between syntactically pseudo-active *kick the bucket* and more active *spill the beans* naturally follows from whether the idiomatic part has a role in the predicate-argument structure, which we capture by systematically choosing between option (4c/i) and (4c/ii) per lexical correspondence.

3.1   Parsing and Correspondence with Singleton Types

The crucial property of a category in a lexical correspondence such as $\alpha :=$ $A/$ "s" with singleton s, is that the string "s" *as a category* does have its own correspondence. This cannot be a literal match without categorial processing of the surface string, with s to the right of α. It is a compositional derivational process arising from (a) below, to lead to (b). The lexically specifiable difference from a polyvalent category such as *NP, VP* is that the item α subcategorizes for the string s, hence treat it as a category, rather than subcategorize for the category of s, viz. B in the example. To obtain B, the derivational process works as usual for s, independent of the item α. We shall see in (9) that rules of function application need no amendment for this interpretation. (8b) is lexically determined by α.

(8)   a. s $:= B: s'$

   b.
$$\frac{\overset{\alpha}{A/\text{``}s\text{''}: \lambda x.p_x} \quad \overset{s}{B: s'}}{\alpha s := A: p_{s'}}>$$

Same idea applies to backward application, for $\alpha := A\backslash$ "s" and the sequence $s\alpha$.

In other words, the surface string s is derived by the derivational process as well. It is just that the item α carrying the singleton type as an argument decides what to do with its semantics, which we indicated schematically above as modal contribution to contingency of p, as p_x of α. This is not post-processing of a category in a radically lexicalized grammar, in which all and only head functors decide what to do with the semantics of their arguments.

It means that, whether an argument type is polyvalent or singleton, there has to be an LF placeholder for it, otherwise the derivational process, which is completely driven by syntactic types in CCG, cannot proceed. It can be seen in the basic primitive of CCG, viz. function application:

(9) $X/Y : f \ Y : a \quad \rightarrow X : fa$ (>)
     $Y : a \quad X\backslash Y : f \rightarrow X : fa$ (<)

The LF of the functor, f, has to be a lambda abstraction, to be able to take any Y and yield fa. This is true of singleton '$/Y$' and '$\backslash Y$' too.

We can clearly see the role of substitution rather than insertion in projection of types. The rule (>) above is in fact realized as below (similarly for others):

(10) $\alpha := X/Y : f \quad \beta := Y : a \quad \rightarrow \quad \alpha\beta := X : fa$ (>)

There is no sense in which we can insert something into α and β as they form $\alpha\beta$ because these are surface expressions.

The singleton types present an asymmetry in argument-result (or domain-range) specification. Functors such as A/B and $A\backslash B$ have domain B and range A, and, apart from trivial identities where A and B are the same singleton, the interpretation where the range itself (A) is a singleton is problematic. Since $A|B$ is a function *into* A for some slash '$|$', if it is not a trivial case of singleton identity, say *"up"*/*"up"*, it is difficult to see how A can be singleton. Although there are no formal reasons to avoid singleton results, and results of results, we conjecture that singletons are arguments, and arguments of results and arguments, because there seems to be no nontrivial function of a singleton-result with grammatical significance.

A related argument can be made about a singleton's potential to be the overall syntactic category of a lexical item. The notion of extending the phonological range of an item such as (a) below coincides naturally with "words with spaces" idea (e.g. *ad hoc, by and large, every which way*), which is common in NLP of MWEs, but (b) is also an option.

(11) a. every which way := $(S\backslash NP)\backslash(S\backslash NP)$: $\lambda p\lambda x.omni'px$

     b. every which way := *"every which way"* : $omniway'$

Notice that (b) is different than having $scored := (S\backslash NP)/$ *"every which way"* for lexically specified verbal adjunction in the manner of [13], which, given (8), must either use entries similar to (11), or derive *every which way* syntactically, and choose to trump its category because it wants a narrower LF due to singleton subcategorization. However we think that both options may be redundant, because of the following.

In CCG the head functor decides the semantics of its entry even if it subcategorizes for a singleton category. Therefore the entries in (a–b) above which we use in (a–b) below *may* be redundant if the words in "words with spaces" are part of the grammar, and if they can combine in any way, say as in (c) below for some A, B, C:

(12) a.
$$\frac{\text{My team} \quad \text{scored} \quad \text{every which way}}{NP \quad (S\backslash NP)/\text{"every which way"} \quad (S\backslash NP)\backslash(S\backslash NP)}$$
$$\overline{\qquad\qquad\qquad S\backslash NP \qquad\qquad\qquad}{}^{>}$$

b.
$$\frac{\text{scored} \qquad\qquad \text{every which way}}{(S\backslash NP)/\text{"every which way"} \quad \text{"every which way"}}$$
$$\overline{\qquad\qquad S\backslash NP \qquad\qquad}{}^{>}$$

c.
$$\frac{\text{scored} \qquad\qquad \text{every which way}}{(S\backslash NP)/\text{"every which way"} \quad \dfrac{A/B \quad \dfrac{B/C \quad C}{B}{}^{>}}{A}{}^{>}}$$
$$\overline{\qquad\qquad\qquad S\backslash NP \qquad\qquad\qquad}{}^{>}$$

There would be no post-processing or reanalysis in these cases; they would be multiple analyses because of redundancy. The transparency of derivation requires that in configurations like (8b) the constituents of the rule applying can themselves be derived.

The rules that allow CCG to rise above function application in projection, composition and substitution also maintain the transparency of the syntactic process, by being oblivious to the nature of argument types in these rules:[4]

(13) $X/Y : f \quad Y/Z : g \rightarrow X/Z : \lambda x.f(gx)$ (> **B**)
$\quad\;\; X/Y/Z : f \; Y/Z : g \rightarrow X/Z : \lambda x.fx(gx)$ (> **S**)

If the result categories are not singletons, as we argued, then the rules above never face a case where Y is a singleton. This means that, since singletons are

---

[4] We show only one directional variant of each rule for brevity. The same idea applies to all variants; see Steedman references for a standard set of rules, and [5] for review of proposals for combinatory extensions.

Bozşahin [5]: Sect. 10 shows that all projection rules of CCG can be packed into one monad to enable monadic computation with just one rule of projection. This is possible because CCG is radically lexicalized in the sense that combinatory rules cannot project anything which is not in the lexicon. What appears to be rule choice when presented as (9/13) becomes dependency passing within monad with one rule of combination.

arguments, meaning they bear a slash, say '$|A$' for some slash '$|$' in $\{\backslash, /\}$, the slash is inherently application-only, equivalently '$|_\star A$' in [3] terminology.[5]

This is corroborated by examples like below where there is no idiomatic reading: (We show the derivation for the hypothetical case where singletons would be allowed to compose. Typing the singleton as '$/_\star$ "the bucket"' eliminates the derivation. The slashes in the paper are harmonic '$\backslash_\diamond$' or '$/_\diamond$' unless stated otherwise.)

(14)
$$
\begin{array}{c}
\underline{\text{♯John kicked}} \quad \underline{\text{and}} \quad \underline{\text{Mary}} \quad \underline{\text{did}} \quad \underline{\text{not}} \quad \underline{\text{kick}} \\
S/\text{"the bucket"} \quad (X\backslash_\star X)/_\star X \quad S/(S\backslash NP) \quad (S\backslash NP)/VP_{\text{inf}} \quad VP_{\text{inf}}/VP_{\text{inf}} \quad VP_{\text{inf}}/\text{"the bucket"} \\
\end{array}
$$

the bucket
NP

For polyvalent types, one-to-one correspondence of syntactic types and placeholder types is meant to capture the thematic structure in CCG, for example for *the door opened* versus *someone opened the door*, by having two different (albeit related) correspondences for *open*.

For a singleton, its functor (and there must be one, since they can only be arguments) decides lexically whether there is a predicate-argument structural role for the placeholder in the LF, as we see in the distinction of *spill the beans*, where *secret'* is an argument of *divulge'*, versus *kick the bucket*, where *bucket'* or anything related to it is not an argument of *die'*.

Therefore, for CCG, MWEs and phrasal idioms are not exceptions that need non-transparent derivation, apart from lexical specification as something special. They are consequences of the nature of categories and radical lexicalization.

Also because of the properties described in this section, a string as a category cannot be empty, which would violate CCG's principle of adjacency and principle of transparency (see Steedman references). No rule in (9) or (13) can apply if one of the categories is empty. Therefore the surface string itself for the singleton (s in example (8)) cannot be empty either.

Having explored the possibilities for the singleton types in combinatory categories, we look at their use.

---

[5] The way this is implemented in many CCG systems including ours is for example to constrain the slashes as follows:

$$X/_\star Y : f \quad Y : a \quad \rightarrow X : fa \qquad (>)$$
$$X/_\diamond Y : f \quad Y/_\diamond Z : g \rightarrow X/_\diamond Z : \lambda x.f(gx) \qquad (>\mathbf{B})$$

It is easier to describe slash-modal control from the perspective of syntactic types of expressions accessing these rules. '$\star$-rules' are accessible by all categories, '$\diamond$' and '$\times$' are compatible only with themselves, and with the most permissive slash.

3.2 Verb-Particles and Phrasal Idioms with Singleton Types

In verb-particle constructions, the differences in the syntax-semantics correspondence force the following lexical distinctions. We now write the categories in more detail than in the preview.

(15)   a. picked := $(S\backslash NP)/$ "up"$/NP_{\text{-heavy}}$: $\lambda y\lambda x\lambda z.cause'(init'(hold_x{}'yz))z$

   b. picked := $(S\backslash NP)/NP_{+\text{lexc}}/$ "up": $\lambda x\lambda y\lambda z.pick_x{}'yz$

   c. picked := $(S\backslash NP)/NP$: $\lambda x\lambda y.pick'xy \wedge choose'xy$

The features above are all finite-state computable, just like morphological ones, as phonological weight ($\mp$heavy) and lexical content ($\mp$lexc) in an expression substituting for a category. All CCG category features can be interpreted this way, because combinators do all the syntactic work.

The reason for having two different grammar entries (a–b) for *pick up* follows from the fact that they are not equally substitutable, for example as an answer to *What did you do?*

(15b) leads to achievement, and (15a) to culmination. Both cases also differ from (c), which provides wider substitution for *NP*, and with a different meaning. We treat (a–c) distinctions surface-compositionally, which are transparently projected without wrap:

(16)

I	picked	the book	up
$\overline{NP_{1s}}$	$\overline{(S\backslash NP)/\text{"up"}/NP_{\text{-heavy}}}$	$\overline{NP}$	$\overline{((S\backslash NP)\backslash(S\backslash NP))/NP}$
$: i'$	$: \lambda y\lambda x\lambda z.cause'(init'(hold'_x yz))z$	$: def'book'$	$: \lambda x\lambda p\lambda y.up'(py)x$

$$\frac{(S\backslash NP)/\text{"up"}}{: \lambda x\lambda z.cause'(init'(hold'_x(def'book')z))z}{\Large{>}}$$

$$\frac{S\backslash NP}{: \lambda z.cause'(init'(hold'_{\lambda x\lambda p\lambda y.up'(py)x}(def'book')z))z}{\Large{>}}$$

$$\frac{S}{: cause'(init'(hold'_{\lambda x\lambda p\lambda y.up'(py)x}(def'book')i'))i'}{\Large{<}}$$

where *hold'* at the end of the derivation can interpret its event modality (contingency) compositionally, since it is a closed lambda term.

Notice that the word *up* knows nothing about the verb-particle construction. Its category is for a PP head, say PP_{up}, as a predicate modifier. It is the verb that delivers the distinct meaning. Its subcategorization is for a singleton, which eschews the syntactic category of the word *up* but not its phonology and semantics, as described in (8b).

(15b) can be assumed to arise from the syntactic category $VP/NP_{+\text{lexc}}/$ "up" by finite inflection. CCG has options here, to accommodate morphology without having to have a "morphological insertion point" in a contiguous but MWE entry *pick up* := $VP/NP_{+\text{lexc}}$, to avoid ?*pick upped*.[6] This is made possible by singleton types.

---

[6] The fact that this form is also attested in child and adult language suggests that these entries may be bonafide lexical options.

Examples (15a–b) use a degree of freedom which is relevant to phrasal idioms. The singleton syntactic type "up" corresponding to the LF placeholder x maintains the compositionality of the correspondence; but, it may have no contribution to the predicate-argument structure at all in some cases, which would make it paracompositional, because its semantic type is a closed lambda term as far as predicate-argument structure is concerned. Notice that in (8b), s' is not in the predicate-argument structure of p; it is a contingency of p.

Consider the following examples in this regard, where x for $bucket'$ as an event modality might mean 'ceremonial death', 'reported death', etc.:

(17)   a. kicked := $(S \backslash NP)/_\star$ "the bucket": $\lambda x \lambda y.die'_x y$

  b. kicked := $(S \backslash NP)/NP$: $\lambda x \lambda y.kick' xy$

They anticipate very limited syntax in the semantically paracompositional part in the idiom reading (a) because of having to enumerate them (*kick the old/proverbial bucket* vs *kick the bucket that John thought overflowed*).[7] These assumptions cannot give rise to the idiom reading in *the bucket that you kicked*, with no further stipulation than singleton categories in a lexical entry (cf. a–b; '*' on the right of a derivation means it is not possible):

(18)   a.   ♯the bucket   that   you   kicked
                        ―――――――――――― ―――――――――― ――――――――――――――――――
                        $(N \backslash N)/(S/NP)$ $S/(S \backslash NP)$ $(S \backslash NP)/_\star$ "the bucket"
                                              ―――――――――――――――――――――――――*>B
                                              $S/$ "the bucket"

  b.   the bucket   that   you   kicked
       ―――― ――   ―――――――――――― ―――――――――― ―――――――――――
       NP/N N $(N \backslash N)/(S/NP)$ $S/(S \backslash NP)$ $(S \backslash NP)/NP$
                                         ――――――――――――――――――――――>B
                                              S/NP
                            ――――――――――――――――――――――――――――――――>
                                     $N \backslash N$
                       ―――<
                                         N
                    ―――>
                                        NP

Given the polyvalent argument category of the relative pronoun, we can see that relativization out of phrasal idioms would not be possible even if we allowed composition of singleton types, therefore the syntactic productivity of idiomatically combining phrases arises from their use of head-dependencies rather than singletons, as we shall soon see in derivations similar to (b), in (26).

We note that carrying the head-word in a polyvalent category to have the same effect, for example $kick := (S \backslash NP)/_\star NP_{\text{bucket}}$, overgenerates the idiom reading, because *the bucket that John thought overflowed* can substitute for NP_{bucket}.

The direct approach to categories that we see in radically lexicalized grammars, whether they are polyvalently substitutable or not, contrasts with systems

―――――――――――

[7] It is tempting to try $NP_{\text{proverbial bucket}}$: $proverb' death'$ for *kick the proverbial bucket* which is a head-subcategorizing category; but, we would have to overextend ourselves to eliminate the idiom reading in *kick the proverbial bucket that overflowed* if we have to. In this sense we suggest that phrasal idioms are best treated with singleton types.

of rewrite and/or record keeping in which post-processing is possible. For example there is no reanalysis or post-processing mechanism needed to eliminate the idiomatic reading below:

(19)   ♯Mary   dragged   and   John   kicked   thebucket.
    $\overline{S/(S\backslash NP)}$ $\overline{(S\backslash NP)/NP}$ $\overline{S/(S\backslash NP)}$ $\overline{(S\backslash NP)/_\star}$ $\overline{\text{``the bucket''}}$
    $\underline{\hspace{5cm}}$ -> B $\underline{\hspace{6cm}}$ -*> B

We can then follow [31] in assuming that passive is a polyvalent lexical process headed by the passive morpheme, mapping for example VP_{inf}/NP to VP_{pass}, which eliminates passivization *the breeze was shot* from the entry:

(20) shoot $:= VP_{inf}/$ *"the breeze"* : $\lambda x \lambda y.smalltalk'_x one'y$

Idioms such as *at any rate, beside the point* further demonstrate that all idioms needing restricted types must contain a predicative element in the domain of locality of their head because we are required by paracompositionality to record the special reading and contingency, for example as extension of discursive clarification (a) and comparison (b):

(21)   a. at $:= (S/S)/$ *"any rate"* : $\lambda x \lambda s.more' exactly' s_x$

    b. at $:= (S\backslash S)/$ *"any rate"* : $\lambda x \lambda s.contrastwith'_x s$

4   Head-Word Subcategorization and Idioms

The difference between idiomatically combining phrases and phrasal idioms such as kicking the bucket is clear: The syntactically active ones are active because the idiomatic part has a role in the predicate-argument structure. 'Secret' is an argument of 'divulge', whereas 'bucket' is not an argument of 'die'. For example, *spill the beans* seems to require categorization such as (a) below in the manner of (6), rather than (b) fashioned from (5) or singleton-subcategorizing (c). Cf. also the non-idiomatic *spill* in (d). Tense morphology renders finite versions of VP_{inf} below as $S\backslash NP$, eg. $spilled := (S\backslash NP)/NP_{beans}$ for (a).

(22)   a. spill $:= VP_{inf}/NP_{beans}: \lambda x \lambda y.divulge'_x secret'y$

    b. spill $:= VP_{inf}/NP$: $\lambda x \lambda y.$if $head(x) = beans'$then $divulge'_x secret'y$
        else $spill'xy$

    c. spill $:= (VP_{inf}/$ *"beans"*$)/PredP: \lambda p \lambda x \lambda y.divulge'_{px} secret'y$

    d. spill $:= VP_{inf}/NP: \lambda x \lambda y.spill'xy$

PredP is a predicative phrase type, which includes the quantifier phrase. The syntactic type of the idiomatic argument in (a) encodes the head-dependency from surface structure. It avoids the idiomatic reading in *to spill the bean*, which (b) may not. (b)-style solutions would depend on LF objects, which may not always reflect surface forms in full. In fact (b) requires post-processing to eliminate the idiom reading in the following example:

(23)

♯You	spilled	and	Mary	cooked	the beans
$S/(S\backslash NP)$	$(S\backslash NP)/NP$	$(X\backslash_{\star}X)/_{\star}X$	$S/(S\backslash NP)$	$(S\backslash NP)/NP$	NP_{beans}
$:\lambda p.p\ you'$	$:\lambda x\lambda y.\text{if}\ \cdots$	$:\lambda p\lambda q\lambda z.\text{and}''(pz)(qz)$	$:\lambda p.p\ m'$	$:\lambda x\lambda y.\text{cook}'xy$	$:\text{def}\ beans'$

$$\frac{}{\begin{array}{c} S/NP \\ :\lambda x.\text{if } head(x) = \cdots \end{array}} >_{\mathbf{B}}$$

$$\frac{}{\begin{array}{c} S/NP \\ :\lambda x.\text{cook}'xm' \end{array}} >_{\mathbf{B}}$$

$$\frac{}{\begin{array}{c} S/NP \\ :\lambda z.\text{and}'(\text{if } head(z) = \cdots)(\text{cook}'z\ m') \end{array}} \&$$

$$\frac{}{\begin{array}{c} S/NP: \text{and}'(\text{if } head(\text{def } beans') = \\ beans' \text{ then } divulge'\ secret'\ you'\ \cdots)(\text{cook}'(\text{def } beans')\ m') \end{array}} >$$

This is still the case if we treat the construction as multi-headed, as [15]:238 do, by also assuming *the beans* $:= NP_{\text{beans}}: secret'$, and changing the LF choice condition of *spill* to 'if $head(x) = secret'$ then $divulge'xy$ else $spill'xy$'. *Cook'* does not refer to this entry.

The process of marking head-word dependencies requires statistical learning, as the category such as NP_{beans} in (22a) implies. It has been known in TAG systems with supertags since [6] that disambiguating such categories is feasible with training. The earliest approach to such marking in CCG is [8,9] as far as we know, where probabilistic CCGs are similarly trained. Later work such as [1] shows further progress in disambiguation of head-dependencies.

NP_{beans} is a polyvalent type, not a singleton. Therefore we get the following accounted for by (22a) (some of the examples are from [33]):

(24)   a. spill /several/the musical/the artistic/mountains of/loads of/ beans

b. spill the beans no one cares about

Head-marking of an argument category by the idiom's head is required because of examples such as below, where an idiomatic reading is eliminated despite relatively free syntax because the coordinands would not be like-typed:

(25)

♯You	spilled	and	Mary	cooked	the beans
$S/(S\backslash NP)$	$(S\backslash NP)/NP_{\text{beans}}$	$(X\backslash_{\star}X)/_{\star}X$	$S/(S\backslash NP)$	$(S\backslash NP)/NP$	NP_{beans}

$$\frac{}{S/NP_{\text{beans}}} >_{\mathbf{B}}$$

$$\frac{}{S/NP} >_{\mathbf{B}}$$

$$\frac{}{} *\&$$

Right-node raising succeeds when non-idiomatic entries such as (22d) do not subcategorize for head-word marked arguments. (25) is unproblematic with it.

When the head of the construction does not require identical types as does the conjunction above, head-projection works with simple term match; cf. the one for kicking the bucket in (18a) (h is for head-word feature):

(26)

the	beans	that	you	spilled
NP_h/N_h	N_{beans}	$(N_h\backslash N_h)/(S/NP)$	$S/(S\backslash NP_{2s})$	$(S\backslash NP)/NP_{\text{beans}}$

$$\frac{}{S/NP_{\text{beans}}} >_{\mathbf{B}}$$

$$\frac{}{N_h\backslash N_h} >$$

$$\frac{}{N_{\text{beans}}} <$$

$$\frac{}{NP_{\text{beans}}} >$$

The example also shows that argument types of idiomatically combining phrases must be composable; therefore; (22c) is inadequate.[8]

5   Idioms Requiring a Combined Approach

There seems to be cases where a combination of singletons and head-marked polyvalent subcategorization is needed. The *give creeps* construction, which is sometimes considered not an idiom because of its compositionality [19], is para-compositional in our sense, and idiomatically combining in [24] terminology, because although *creeps* seems to be an event modality of *revulse'* rather than its argument, *fear'* is an argument. A simple head-marking approach such as '$/NP_{\text{creeps}}$' would overgenerate in cases such as ♯*give me some creeps*, but we have *give me the absolute/shivering/full-on creeps*. Notice also that the construction and related items resist dative shift (judgments are from [20]; '*' seems to be equivalent to '♯' in our terms):

(27)   a. The Count gave me the creeps./ *The Count gave the creeps to me.

     b. His boss gave Max the boot./ *His boss gave the boot to Max.

Richards [26] observes that (a) below can be the unaccusative of *give*; and, (b) is widely attested in the web (but recall ♯*give me some creeps*).

(28)   a. Mary got the creeps.

     b. give some creeps

     c. give := $VP/N_{\text{creeps}}/$ "*the*"$/NP$: $\lambda x\lambda y\lambda z\lambda w.cause'(init'(revulse'_z fear'_y x))w$

Assuming that dative shift is polyvalent, following [31], in the form of lexical mapping from $VP/NP/NP$ to $VP/PP_{\text{to}}/NP$, we can eliminate it for the type in (c), which we think captures the insight of Richards, and permits adjunction within an N, e.g. *mountains of creeps*.

Another class of idioms forces a combined approach as well. Semantic reflexives in *I twiddled my thumbs/ate my words/racked my brain/lose my mind* are not morphological reflexives and they are inherently possessive, for example:

(29) twiddled := $(S\backslash NP_{\text{agr}})/$ "*thumbs*"$/NP_{\text{-lexc,+poss,agr}}$
$\qquad\qquad\qquad : \lambda x\lambda y\lambda z.pass'_y\ time'_{(self'_z)}z \wedge inalien'(xyz)$

---

[8] One way to put it altogether is to use a feature such as ∓special in addition to h, which ordinary verbs negatively specify, heads of idiomatic combination positively specify, and heads of syntactic constructions eg. coordinators and relative markers (under)specify as they see fit. The value '+special' need not be further broken down for singletons because they are self-representing, and, presumably, featureless. For example phonological weight is intrinsically captured in "*the beans*"; also, lexical content.

The LF captures the properties that the subject idles on his own time, the lexical possessive in the LF of x which is presumably lexically *poss′* is inalienable and belongs to the subject. This is a reflexive in the sense that it must be bound in its local domain determined by *pass′*. The referent (z) is available in one domain of locality in a radically lexicalized grammar because the head of the idiom does not require a VP in phrase-structure sense but a clause. Agreement is locally available too; by insisting on same agreement features. The head-dependency is that the argument does not contain lexical material, leaving out examples such as *John twiddled John's thumbs* as an idiom.

6   No Wrap

We have seen that options (4c/i) and (4c/ii) are not mutually exclusive. We also suggested that singleton type is a forced move to avoid loss of meaning composition. One consequence of this is the treatment of verb-particles without wrap, which are not related to idioms although they are MWEs. We now consider option (4b) in more detail from this perspective, which at first sight seems to be just as lexical as the two alternatives we have considered so far.

The projection principle of CCG, which says that lexical specification of directionality and order of combination cannot be overridden during derivations, eliminates (30) from projection because it has the second-combining argument (Y) of a function applying before its first-combining argument (Z), an operation of the general class that has been proposed in other categorial approaches under the name of "wrap."

(30) $(X/Y)/Z: f \ Y: a \rightarrow X/Z: \lambda z.fza$ (*)

Wrap of the kind in (30) has a combinatory equivalent, namely Curry's combinator **C** (see [11]). CCG's adjacency principle eliminates this combinator on empirical grounds, rather than formal, as a freely operating rule. Adding (30) to CCG's projection has the effect of treating VSO and VOS as both grammatical, which is not the case for Welsh, and to carry the same meaning, which is not the case for Tagalog although both VSO and VOS are fine. These properties must be part of a lexicalized grammar rather than syntactic projection.

The version of wrap which [2,12,16] employ is different, which was eliminated from consideration so far because it is non-combinatory; and, it violates adjacency of functors and arguments. That wrap is the following:

(31)
$$\frac{s_1 \qquad\qquad s_2}{\dfrac{X/_W\,Y: f \qquad Y: a}{first(s_1)\,s_2\,rest(s_1) := X: fa}} \text{wrap}$$

where $first()$ function gives the first element in a list of surface expressions for Bach [2], or first word for Dowty [12]; and, $rest()$ returns the rest of the expression. The wrapping slash '$/_W$' of Jacobson [16] does the infixation of s_2.

Semantically, it is function application. Syntactically, no combinator can do what this rule does to its input expressions, which is to rip apart one surface expression (s_1) and insert into it. It differs from **C**, which wraps one independent expression in *two* independent expressions.

The appeal of surface wrap to MWEs was to be able to write a category for *pick* ⋯ *up* as for example $pick := (S\backslash NP)/_W NP/P_{up}: \lambda x\lambda y\lambda z.pick_x{'}yz$; cf. (16).

Syntactic wraps such as above, whether combinatory or non-combinatory, have domino effects on dependency and constituency, unlike 'lexical wrap', where a lexical entry specifies its correspondence; for example, for the strictly VSO Welsh verb $gwelodd := (S/NP)/NP_{3s}: \lambda x\lambda y.saw{'}yx$; note the LF.

An example of global complications in grammar caused by wrap can be seen below, where dashed boxes denote wrapped-in material; cf. Figure 1.

(32) a.

$$\frac{\underline{\text{persuade}} \qquad \underline{\text{to do the dishes}} \qquad \underline{\text{John}}}{VP_{inf}/_W NP/VP_{inf} \qquad\qquad VP_{inf} \qquad\qquad NP}$$

$$\cfrac{\boxed{\text{persuade} \big\| \text{to do the dishes}} := VP_{inf}/_W NP}{\boxed{\text{persuade} \big| \text{John} \big| \text{to do the dishes}} := VP_{inf}} \text{{\scriptsize wrap}}$$

b.

$$\frac{\underline{\text{persuade}} \qquad \underline{\text{to do the dishes}} \qquad \underline{\text{John easily}}}{VP_{inf}/_W NP/VP_{inf} \qquad\qquad VP_{inf} \qquad\qquad NP}$$

$$\boxed{\text{persuade} \big\| \text{to do the dishes}} := VP_{inf}/_W NP$$

$$\boxed{\text{persuade} \big| \text{John} \big| \text{to do the dishes}} := VP_{inf} \quad\text{{\scriptsize wrap}}$$

$$\boxed{\text{persuade John to do the dishes} \big\| \text{easily}} := VP_{inf}$$

c.

$$\frac{\underline{\text{persuade}} \qquad \underline{\text{John}} \quad \underline{\text{to do the dishes easily}}}{VP_{inf}/VP_{inf}/NP \quad NP \qquad\quad VP_{inf}}$$

$$VP_{inf}/VP_{inf}$$

Derivation (a) is Bach's use of non-combinatory wrap rule in (31). Given these categories which involve wrap, there is one interpretation for (b), where the adverb can only modify *persuade*. With the unwrapped version of *persuade* in (c), two interpretations are possible: one modifies the VP complement of *persuade*, and the other, *persuade John*, both of which are required for adequacy.

7   Conclusion

One point of departure of CCG from other categorial grammars and from tree-rewrite systems is that (i) we can complicate the basic vocabulary of the theory, but (ii) not its basic mechanism such as introducing wrap, if a better explanation can be achieved. The first point has been made by Chomsky repeatedly since [7]:68. Singleton types could be viewed as one way of doing that, much like $S\backslash NP$

vs. *VP* distinction. We have argued that it is actually not a complication at all in CCG's case, because the possibility has been available, in the notion of type as a set of values, which can be a singleton set. CCG differs from Chomskyan notion of category substitution by eliminating *move*, empty categories and lexical insertion altogether, which means that all computation is local, type-driven, and there is no action-at-a-distance, to address the second point. The expressions substituting for these types are then locally available in the course of a derivation. This seems critical for MWEs.

The possibility of a singleton value is built-in to any type. The asymmetry of CCG's singletons' categorization, that they can be arguments, and arguments of arguments and results, and, their inherent applicative nature, deliver MWEs and phrasal idioms as natural consequences rather than stipulation or a "pain in the neck for NLP." Syntactically active idioms are not singleton-typed because they have relevance to predicate-argument structure; and, their narrower syntax, compared to free syntax, seems to necessitate head-marking of some argument categories, which is known to be probabilistically learnable.

Some implications of our analyses are that all idioms can be made compositional at the level of a lexical correspondence without losing semantic distinctions, and without meaning postulates or reanalysis. Categorial post-processing of MWEs and phrasal idioms, and multi-stage processing of them in the lexicon, as done by [10,33], may be unnecessary if we assume type substitution to be potentially having one value, and surface head-marking to be an option for polyvalent argument types. One conjecture is that any idiom in any language has to involve a predicate implicated by some predicative element in the expression to keep the meaning assembly paracompositional.

The analyses in the article can be replicated by running the CCG tool at github.com/bozsahin/ccglab. The particular fragment in the chapter is at github.com/bozsahin/ccglab-grammars/cb-ag-fg2018-grammar.

References

1. Artzi, Y., Lee, K., Zettlemoyer, L.: Broad-coverage CCG semantic parsing with AMR. In: Proceedings of the 2015 Conference on Empirical Methods in Natural Language Processing, Lisbon, Portugal, pp. 1699–1710 (2015)
2. Bach, E.: An extension of classical transformational grammar. In: Problems in Linguistic Metatheory: Proceedings of the 1976 Conference at Michigan State University, Lansing, MI, Michigan State University, pp. 183–224 (1976)
3. Baldridge, J., Kruijff, G.J.: Multi-modal combinatory categorial grammar. In: Proceedings of 11th Annual Meeting of the European Association for Computational Linguistics, Budapest, pp. 211–218 (2003)
4. Bond, F., Ho, J.Q., Flickinger, D.: Feeling our way to an analysis of English possessed idioms. In: Müller, S. (ed.) Proceedings of the 22nd International Conference on Head-Driven Phrase Structure Grammar, Stanford, CA, CSLI Publications, pp. 61–74 (2015)
5. Bozşahin, C.: Combinatory Linguistics. De Gruyter Mouton, Berlin (2012)
6. Chen, J., Bangalore, S., Vijay-Shanker, K.: New models for improving supertag disambiguation. In: Proceedings of the Ninth Conference on European Chapter

of the Association for Computational Linguistics, pp. 188–195. Association for Computational Linguistics (1999)

7. Chomsky, N.: Some empirical issues in the theory of transformational grammar. In: Peters, S. (ed.) Goals of Linguistic Theory. Prentice-Hall, Englewood Cliffs (1972)

8. Clark, S., Curran, J.R.: Wide-coverage efficient statistical parsing with CCG and log-linear models. Comput. Linguist. **33**(4), 493–552 (2007)

9. Clark, S., Hockenmaier, J., Steedman, M.: Building deep dependency structures with a wide-coverage CCG parser. In: Proceedings of the 40th Annual Meeting on Association for Computational Linguistics, pp. 327–334 (2002)

10. Copestake, A.: Representing idioms. In: HPSG Conference, Copenhagen (1994)

11. Curry, H.B., Feys, R.: Combinatory Logic. North-Holland, Amsterdam (1958)

12. Dowty, D.: Dative movement and Thomason's extensions of Montague grammar. In: Davis, S., Mithun, M. (eds.) Linguistics, Philosophy, and Montague Grammar, pp. 153–222. University of Texas Press, Austin (1979)

13. Dowty, D.: The dual analysis of adjuncts/complements in categorial grammar, pp. 33–66 (2003). in [18]

14. Everaert, M., Van der Linden, E.J., Schreuder, R., Schenk, A. (eds.): Idioms: Structural and Psychological Perspectives. Lawrence Erlbaum, Mahwah (1995)

15. Gazdar, G., Klein, E., Pullum, G., Sag, I.: Generalized Phrase Structure Grammar. Harvard University Press, Cambridge (1985)

16. Jacobson, P.: Flexible categorial grammars: Questions and prospects. In: Levine, R. (ed.) Formal Grammar, pp. 129–167. Oxford University Press, Oxford (1992)

17. Kay, P., Sag, I.A., Flickinger, D.: A lexical theory of phrasal idioms. www1.icsi.berkeley.edu/~kay/idiom-pdflatex.11-13-15.pdf (ms)

18. Lang, E., Maienborn, C., Fabricius-Hansen, C.: Modifying Adjuncts. Walter de Gruyter, Berlin (2003)

19. Larson, R.: On the double object construction. Linguist. Inquiry **19**, 335–392 (1988)

20. Larson, R.: On "dative idioms" in English. In: Workshop on Syntax-Semantics. Fuji Women's University (2012)

21. Lichte, T., Kallmeyer, L.: Same syntax, different semantics: a compositional approach to idiomaticity in multi-word expressions. In: Piñón, C. (ed.) Empirical Issues in Syntax and Semantics, vol. 11, pp. 111–140. CSSP, Paris (2016)

22. Manning, C.D.: Ergativity: Argument Structure and Grammatical Relations. CSLI, Stanford (1996)

23. Moens, M., Steedman, M.: Temporal ontology and temporal reference. Comput. Linguist. **14**, 15–28 (1988). Reprinted in Mani, I., Pustejovsky, J., Gaizauskas, R. (eds.) The Language of Time: A Reader, pp. 93–114. Oxford University Press

24. Nunberg, G., Sag, I.A., Wasow, T.: Idioms. Language **70**(3), 491–538 (1994)

25. Pulman, S.G.: The recognition and interpretation of idioms. In: Cacciari, C., Tabossi, P. (eds.) Idioms: Processing, Structure, and Interpretation, pp. 249–270. Lawrence Erlbaum, Hillsdale (1993)

26. Richards, N.: An idiomatic argument for lexical decomposition. Linguist. Inquiry **32**(1), 183–192 (2001)

27. Sag, I.A., Baldwin, T., Bond, F., Copestake, A., Flickinger, D.: Multiword expressions: a pain in the neck for NLP. In: Gelbukh, A. (ed.) CICLing 2002. LNCS, vol. 2276, pp. 1–15. Springer, Heidelberg (2002). https://doi.org/10.1007/3-540-45715-1_1

28. Schuler, W., Joshi, A.: Tree-rewriting models of multi-word expressions. In: Proceedings of the ACL Workshop on Multiword Expressions: from Parsing and Generation to the Real World, Portland, OR, ACL, pp. 25–30 (2011)

29. Steedman, M.: The Syntactic Process. MIT Press, Cambridge (2000)
30. Steedman, M.: Taking Scope: The Natural Semantics of Quantifiers. MIT Press, Cambridge (2012)
31. Steedman, M., Baldridge, J.: Combinatory categorial grammar. In: Borsley, R., Börjars, K. (eds.) Non-Transformational Syntax, pp. 181–224. Blackwell, Oxford (2011)
32. van der Linden, E.J.: Incremental processing and the hierarchical lexicon. Comput. Linguist. **18**(2), 219–238 (1992)
33. Villavicencio, A., Copestake, A., Waldron, B., Lambeau, F.: Lexical encoding of MWEs. In: Proceedings of the Workshop on Multiword Expressions: Integrating Processing, pp. 80–87. Association for Computational Linguistics (2004)

Case Theory in Minimalist Grammars

Sabine Laszakovits[(⊠)]

Department of Linguistics, University of Connecticut, Storrs, USA
sabine.laszakovits@uconn.edu

Abstract. This paper investigates the consequences of one-to-many licensing relationships for Minimalist Grammars (MGs; [30]) on the example of case. Dependent Case Theory [2,23] has proposed that a single noun phrase can assign accusative case to arbitrarily many other noun phrases in particular structural configurations. Taking a licensing view rather than an assignment view on the distribution of case, this implies that accusative case can be licensed by a single licensor on arbitrarily many licensees. This paper argues that the distribution rules for case can be formalized as at most monadic second-order constraints, which are known to be translatable into an MG with refined Merge-features [16]. However, an implementation as Move-features is not feasible because such an MG would need to "count" and would thereby generate non-regular derivation tree languages. It is argued that this increase in complexity can be avoided by suspending the SMC for licensing relationships that involve neither displacement of phonological nor of semantic features.

Keywords: Dependent Case Theory · Minimalist Grammar
Licensing · Persistent features · SMC

1 Introduction

In natural language syntax, long-distance dependencies are sometimes formalized as covert movement. The licensee moves into the licensor's specifier position and establishes a local feature checking/valuation relationship without effects on the word order at PF, nor on the scope relations at LF. In some constructions, one licensor is involved in multiple dependencies of the same type, even in arbitrarily many. This paper explores one such construction in detail: accusative assignment in the framework of Dependent Case Theory (DCT). DCT argues that accusative is assigned to a noun phrase (NP) in the presence of another NP obeying certain structural configurations. Phrased differently, one NP can license accusative case on arbitrarily many NPs standing in these configurations. Other examples of one-to-many licensing relationships in natural language syntax include NPI-licensing, anaphor-binding, negative concord, agreement, parasitic morphology, and sequence of tense.

The questions that arise are the following:

© Springer-Verlag GmbH Germany, part of Springer Nature 2018
A. Foret et al. (Eds.): FG 2018, LNCS 10950, pp. 37–61, 2018.
https://doi.org/10.1007/978-3-662-57784-4_3

1. What is the state of the derivation at the point of applying these movement operations?
   How does the derivation keep track of the arbitrary number of licensees?
2. What is the decision procedure to determine in what order the licensees move?

I will show that one-to-many licensing configurations cannot be computed within the power usually assumed for natural languages (regular tree languages; Sect. 4). We do not have independent reason to justify the increase in power to context-free tree languages, which is necessary to capture these arbitrarily many movement steps. In fact, we can show that a formalization within regularity *is* possible as long as we do not resort to covert movement to formalize it, but employ MSO-definable constraints imposed by refined selection features [16] (Sect. 3). This illustrates a fundamental asymmetry between selection and licensing: every constraint that is definable by MSO-logic can be expressed by selection, but not necessarily by licensing.

Minimalist Grammars [30] have formalized the restriction to regularity of their derivation trees into what is known as the Shortest Move Constraint (SMC)—a categoric constraint against arbitrarily many licensees simultaneously awaiting movement [29]. This paper exhibits a formal procedure to circumvent the SMC for particular feature checking relationships without increasing the necessary power of the formalism (Sect. 5). Given that the licensing movement exhibited by DCT influences neither PF nor LF, we can avoid answering question 2. Rather, we can establish all required licensing relationships simultaneously, irrespective of their number. Thus the formalism does not need to remember the number of waiting licensees, answering question 1. At any given time, the only relevant information is whether or not *at least one* licensee is in need of licensing. Whenever a licensing relationship is established, *all* unchecked licensees will be checked.

This then establishes a way to formalize one-to-many licensing relationships with movement features in frameworks that hitherto were constrained by the SMC.

Predictions for wh-Movement. Constructions with one-to-many licensing relationships requiring movement that influences PF have been discussed in the literature on the example of multiple wh-fronting [12,13]. This phenomenon is outside the scope of this paper, but I do make a prediction about it: For constructions in which not all fronted wh-words stand in a c-command relation to another fronted wh-word (in their respective base positions), and not increasing the formalism's power, we predict that the order of moved wh-phrases is not determined by the derivation, i.e., it is either completely free (arbitrary) or completely fixed (determined by a deterministic post-syntactic mapping establishing linear order without reference to syntactic features). However, see [5] and references cited therein on the absence of wh-movement as well as wh-movement to lower positions.

1.1 Outline

This paper is structured as follows. In Sect. 1.2, I introduce the basic idea behind Dependent Case Theory and illustrate the challenges raised by unbounded dependent-down case assignment. Section 2 defines Minimalist Grammars as regular tree grammars generating derivation tree languages [18,30,32]. The following three sections discuss possible implementations of MGs generating tree languages whose distribution of accusative case matches Dependent Case Theory. Section 3 proves that an implementation with a regular tree language is possible by giving an implementation using monadic second-order constraints. Section 4 shows that an implementation with long-distance licensing relationships is not possible without significantly increasing the complexity class of this tree language from a regular to a context-free tree language. Section 5 suggests an amendment: long-distance licensing relationships can be employed if we adapt the formalism so that dependent-case is only assigned to a single nominal, rather than to unboundedly many. Section 6 concludes.

1.2 Dependent Case

Dependent Case Theory (DCT; [23,34]) regulates how and when nominals receive morphological case-marking. At present, the rules that generate the distribution of case morphology in various works on DCT [2,4,21,23,27,28] are given as high-level descriptions of the relevant configurations. For each case there are rules specifying the contexts in which it can be assigned. These rules fall into four categories, defining *inherent cases*, *dependent cases*, *unmarked cases*, and *default cases*, which in turn stand in a hierarchy (1) such that if a noun is eligible for more than one case, only the case in the highest-ranked (left-most) category will be assigned.

(1)   inherent cases > dependent cases > unmarked cases > default cases [23]

Inherent cases (also: lexical, quirky, idiosyncratic cases) are assigned due to a fixed property of the element that introduces the nominal into the derivation. For example, in Icelandic the verb 'to help' always assigns dative to its helpee-argument, even under passivization, where non-inherently accusative marked arguments undergo case alternation to nominative. We use the term *structural cases* to refer to all non-inherent cases.

Dependent cases are structural cases that are assigned in configurations of two or more nominals in the same case assignment domain that stand in a c-command relationship. Dependent-down cases are assigned to the c-commanded nominal, and dependent-up cases to the c-commanding nominal. For example, accusative is a dependent-down case and applies in the CP domain. Dependent-up cases include ergative [3] in the CP domain, and dative in the VP domain [4].[1]

Unmarked cases arise when dependent cases do not, i.e., in the absence of other elements in the structure, and they are sensitive to the case assignment domain. In most languages, nominative is the unmarked case in the CP domain. Genitive has been argued to be the unmarked case in the DP domain [2,23].

Default cases arise outside of case assignment domains, such as in fragment answers or hanging topics. In many languages, default case is nominative, however, for English it has been argued that default case is accusative [23].

Algorithm. An algorithm based on Marantz's [23] disjunctive case hierarchy, in more modern syntactic terminology of e.g. [2], is given in Fig. 1. As soon as the noun phrase (NP) α receives case, the algorithm is exited, leading to at most one case per NP per case-assignment domain. Together with the algorithm's assertion that every NP will receive case[2], this leads to an assignment of *exactly* one case per NP per case-assignment domain. The rule ordering of dependent-down case before dependent-up case captures the fact that in a sequence of

Precondition: α does not have case.
1. The lexical item that merges α can assign *inherent case* to α.
2. If α is still case-less, determine the case-assignment domain D that α is in.
    2.1. If there is another NP β in D that does not have inherent case, and β c-commands α, then assign to α the *dependent-down case* associated with D (if any).
    2.2. If α is still case-less and there is another NP γ in D that does not have inherent case, and α c-commands γ, then assign to α the *dependent-up case* associated with D (if any).
    2.3. If α is still case-less, assign it the *unmarked case* associated with D.
3. If α is still case-less (i.e., it is not in a case-assignment domain), assign it *default case*.
Postcondition: α has case.

Fig. 1. Algorithm for case assignment in DCT

---

[1] Language differ with respect to restrictions on the licensing nominals, in particular with respect to their case marking. In many languages, nominals with inherent case cannot be licensors ("quirky subjects", e.g. in Icelandic [34], Diyari, Kannada [2]). If the subject carries dative, the object will carry nominative, not accusative. However, Tamil as well as some dialect of Faroese exhibit DAT–ACC patterns [2, pp. 187–194], and some dialects of Kurdish allow ERG-marking on the subject if the object carries inherent DAT [1], as is also found in Warlpiri, Burushaski, and Ingush [2, pp. 187–194]. In this paper, I will model the Icelandic patterns, but everything I say extends straight-forwardly to the Faroese pattern.

[2] It has been argued [20, 21] that unmarked cases are the morphological marking that arises in the absence of case. In this paper, I take unmarked cases to be assigned and licensed like other cases.

three NPs α, β, γ such that α c-commands β, and β c-commands γ, all in the same domain D associated with both a dependent-up case (e.g., ergative) and a dependent-down case (e.g., accusative), NP β receives accusative rather than ergative; [2, p. 232] on Diyari.

Dependent-Down Case Licensing is Unbounded. Dependent-down case licensing has a property that the other categories of case do not share. A single NP can assign dependent-down case to unboundedly many nominals in its c-command domain. In (2), arrows indicate assignment of accusative case.

(2)   NP$_1$ NP$_2$ NP$_3$ NP$_4$ $\cdots$

In principle, there are two kinds of configurations for such a series of NPs to stand in. The first is pairwise c-command relations: NP$_1$ c-commands NP$_2$, NP$_2$ c-commands NP$_3$, NP$_3$ c-commands NP$_4$ etc. Asymmetrical c-command is transitive: it follows that NP$_1$ also c-commands NP$_3$ and NP$_4$, that NP$_2$ also c-commands NP$_4$ etc. We thus have multiple potential case assigners for NP$_i, i > 2$: NP$_1$, ..., NP$_{i-1}$. We will refer to this configuration as *daisy-chain* configuration.

The second configuration contains NP$_2$ to NP$_n$ in such positions that for all i: NP$_i$ does not c-command any NP in the above sequence. Thus all dependent-down case-marked NPs in this sequence only have one potential case assigner: NP$_1$. We will refer to this construction as *1:n*-configuration. Clearly, combinations of these two configurations are also possible.

The daisy-chain configuration is well-attested, for example in Finnish [22, 26]. In a sequence like (2), we can observe that removal of NP$_1$ results in NP$_2$ changing its case from accusative to nominative, but all following NPs retain accusative case. Similarly, additional removal of NP$_2$ changes the case on NP$_3$ from accusative to nominative, but the case on NP$_4$ etc. remains unchanged. Under the assumption that the structures with NP$_1$ and without NP$_1$ are identical in all relevant aspects except for the presence or absence of NP$_1$, this is evidence for NP$_{i+1}$ receiving its accusative case from NP$_i, i > 1$, and not from NP$_1$.[3] The daisy-chain configuration is directly expressable by regular tree grammars using licensor features for case and thereby does not pose a problem for the formalism.

The 1:n-configuration is attested in Sakha [4, 33]. In "raising to object" constructions, the subject of an embedded clause (finite or nominalized) can receive accusative marking.

---

[3] [20, 21] argue that only case-less NPs (carrying unmarked case) can license dependent case. If this is correct, the daisy-chain configuration for case assignment does not exist, and these structures employ the 1:n-configuration instead where all assignment is performed by NP$_1$.

(3)   *Masha*   [*Misha-ny   kel-ie   dien*] *djie-ni   xomuj-da.*
       Masha.NOM   Misha-ACC come-FUT.3SG that   house-ACC tidy-PST.3SG
       'Masha tidied up the house (thinking) that Misha would come.'   [33, 368]

Baker and Vinokurova [4, Sect. 3.5] show that the availability of this accusative
marking is dependent on the presence of a matrix subject, and not on other
sources of accusative marking (such as functional heads) in either the embedded
clause or the matrix clause. When changing the matrix predicate to subject-less
predicates like 'it became certain' or 'it became necessary', the embedded subject
cannot carry accusative, and the matrix object cannot carry accusative.[4] This
then supports an analysis like (4), where the source of both accusatives is the
matrix subject.

(4)

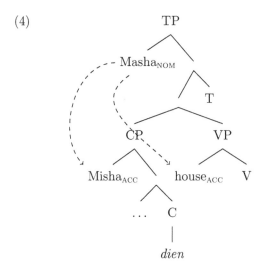

There remains the empirical question whether Sakha allows arbitrarily many
instances of raising to object in one sentence. Given that *dien*-clauses are probably adjuncts, it is not unreasonable to think that there could be more than one
in a single clause. Such a configuration is illustrated in (5).

---

[4] They do not, however, provide an example showing that both NPs lose their
accusative-marking together.

(5)

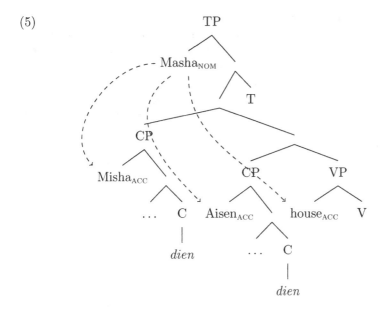

This paper does not aim to contribute to this empirical question and will instead assume DCT as formalized in the literature: (4) extends to configurations with unboundedly many adjunct-clauses, whose subject can receive accusative case that is licensed by the matrix subject.

A Similar Issue with Unmarked Case. Similar to dependent-down cases, a single clause can contain unboundedly many NPs carrying unmarked case. Unlike dependent-down cases, unmarked cases do not have an obvious licensor, however they are dependent on the case-assignment domain. Formally, we can interpret this as the head of this domain (for [2]: the phase head) licensing unmarked case. Then the treatment of unmarked case becomes parallel to that of dependent-down case. This is illustrated in (6), after a construction discussed in [2, p. 86]: Amharic dyadic unaccusatives (possessor or experiencer constructions) as well as passives of triadic verbs with a goal-argument. Since the two NPs in (6) do not stand in a c-command relationship, both are realized with unmarked case.

(6)

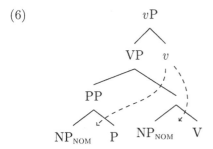

Case as Agree. An alternative option arises from feature valuation and Late Insertion: unmarked (and also default) cases are the elsewhere-forms of nominals that arise when lexical and dependent cases are not available. Instead of having nominals with unmarked case enter a licensing relationship with the head of the case assignment domain, we can make use of this relationship indirectly. Ermolaeva [11] has established a formal framework for Agree operations. Feature values can be transmitted along branches in a tree, 'riding' on top of merge- and move-operations. Nominals will carry a feature for the unmarked case in their case assignment domain, to be valued indirectly by the domain head. Following [2], i.a., I take case assignment domains to correspond to the complement of phase heads. A C-head will value the features on its complement as nominative (7a), and a D-head as genitive (7b). The outgoing information is specified as a superscript on =T resp. =N together with a right arrow.

(7)   a. $[\; \varepsilon \;::\; \texttt{=T}^{[U:\text{NOM}]} \rightarrow \texttt{C} \;]$ b. $[\; \varepsilon \;::\; \texttt{=N}^{[U:\text{GEN}]} \rightarrow \texttt{D} \;]$

We specify all lexical items to percolate this morphological information down the tree, as illustrated in (8). Every lexical item will receive the value for unmarked case from its selector Z with a left arrow and pass it on to all LIs it selects: =X and =Y in (8). Simplifying [11], (8) uses the variable α to indicate that the value received via Z is the same as the value outgoing via X and Y.

(8) $\left[\begin{array}{c} \phi \\ U : \alpha \end{array} \right] \;::\; \texttt{=X}^{[U:\alpha]} \rightarrow \texttt{=Y}^{[U:\alpha]} \rightarrow \texttt{Z}_{\leftarrow} \;]$

NPs will receive the value for U from their selector as well. They are the only LIs whose phonological realization is sensitive to the value of U. If they have received NOM as value of U, they are pronounced as such, if they have received GEN as genitive, and if U has not been valued, as default case. Since values are transmitted on top of Merge and Move operations, we can straightforwardly use Agree to assign a value for inherent cases, which are selected for directly. However, dependent cases are transmitted at a distance, so they require discussion.[5] The valuation for dependent cases either has to proceed via Move-features, which, as we will see, constitutes a problem, or via refined Merge-features that are sensitive to the presence of the licensor. We will see in Sect. 3 details about such a refinement for a licensing account.

An alternative might be to transmit the hypothetical value D for dependent case from the phase head across the entire case assignment domain, like unmarked case above, and then to introduce a spell-out rule that states that if an NP α has an admissible c-commander, it will spell out its hypothetical dependent case D, and if it doesn't, its hypothetical unmarked case U. I leave an investigation into the possible advantages of either approach for future research. Either way, if unmarked cases are a 1:n-challenge in the same way as dependent-down cases, the present treatment of dependent-down cases will extend to them.

---

[5] [28, p. 205] brings up the idea that dependent cases are feature values that are assigned in the presence of the licensing NP, but does not spell out how this might proceed.

On the other hand, if unmarked cases can be solved by different means, the present treatment of dependent-down cases is not affected.

2 Preliminaries

Our data structure are labeled trees $t = \langle V, \trianglelefteq, label \rangle$ over a ranked alphabet $\langle \Sigma, Arity \rangle$ such that V is a set of nodes (vertices), $\trianglelefteq$ (*dominance*) is a partial order on V, from which derive the predicates *strict dominance* $\triangleleft$ (for all $u, v \in V : u \triangleleft v$ iff $u \trianglelefteq v$ and $\exists w \in V : u \trianglelefteq w$ but not $v \trianglelefteq w$) and *immediate dominance* $\blacktriangleleft$ (for all $u, v \in V : u \blacktriangleleft v$ iff $u \triangleleft v$ and $\not\exists w \in V : u \triangleleft w$ and $w \triangleleft v$), and *label* is a function $V \rightarrow \Sigma$ such that for all $v \in V, \sigma \in \Sigma$: $label(v) = \sigma$ only if the number of nodes that v immediately dominates $|\{v' \in V : v \blacktriangleleft v'\}|$ is equal to $Arity(\sigma)$. Two additional conditions must hold for t: the root condition: $\exists r \in V : \forall v \in V : r \trianglelefteq v$, and the chain condition: $\forall u, v, w \in V : u \trianglelefteq w$ and $v \trianglelefteq w$ implies that either $u \trianglelefteq v$ or $v \trianglelefteq u$ [17].

Following [18], we define Minimalist Grammars (MGs) as regular tree grammars G that generate derivation tree languages MDTL(G), which are subsequently mapped to various output structures by using monadic second-order transductions to specify final landing sites of moved elements, ordering of siblings, and headedness [16, Sect. 1.2.3]. For instance, they can be mapped to phrase structure trees, multi-dominance trees, strings, and many other representations.

We describe MGs by specifying their non-branching terminal nodes (their *lexicon Lex*) and the distribution of their branching terminal nodes with respect to *Lex*. A lexicon *Lex* is a finite set of triples $l = \langle p, s, f_1 \cdots f_n \rangle$ (*lexical items*, LIs) containing phonological content p (not to be specified further), semantic content s (not to be specified further), and a finite sequence $f_1 \cdots f_n$ of syntactic features, $n \geq 1$. We write lexical items as $[\; p :: f_1 \cdots f_n \;]$, omitting the semantic content, with $exponent(l) = p$ and $features(l) = f_1 \cdots f_n$. A syntactic feature f has a type τ and a polarity π. Its type τ classifies f as either selection feature $\tau \in \mathbf{sel}$ or licensing feature $\tau \in \mathbf{lic}$, thus $\mathbf{sel}$ and $\mathbf{lic}$ are disjoint. Its polarity π is either positive or negative. The set *Feat* of syntactic features of a lexicon *Lex* is $Feat = \{\langle \tau, \pi \rangle : \langle \tau, \pi \rangle \in features(l), l \in Lex, \pi \in \{+, -\}\}$. For a syntactic feature $f = \langle \tau, \pi \rangle$, we use the following notations and designations:

	$\pi = +$	$\pi = -$
$\tau \in \mathbf{sel}$	=τ, "selector"	τ, "category"
$\tau \in \mathbf{lic}$	+τ, "licensor"	-τ, "licensee"

The sequence of syntactic features on a lexical item is always such that (i) all positive features precede all negative features, (ii) there is exactly one negative selection feature, (iii) the negative selection feature precedes all other negative features [14, 19].

A derivation tree $t = \langle V, \trianglelefteq, label \rangle$ will be in MDTL(G) depending on the syntactic features on its lexical items (its leaves), the distribution and labels of interior nodes, and well-formedness condition on its root. Nodes labeled Merge and Move introduce associations between two syntactic features of the same type but opposite polarity on two different lexical items. A Merge node m is *projected* by a feature $\texttt{=f}$ on a lexical item l_1 dominated by one of its two children and is an *occurrence* of a feature $\texttt{f}$ of the same type but opposite polarity on a lexical item l_2 dominated by its other child. A Move node m is projected by a feature $\texttt{+g}$ on a lexical item l_1 dominated by its single child c and is an occurrence of a feature $\texttt{-g}$ of the same type but opposite polarity on a lexical item l_2 also dominated by c. There are locality restrictions. Informally, the positive features on an LI must project in order (i.e., for m_i the interior node projected by f_i, $m_{i+1} \triangleleft m_i$) and negative features must find the closest matching occurrence.

The well-formedness conditions on the root of t state that all features on all lexical items in t must have been associated with an interior node, except for one category feature $\texttt{z}$ which must be in a given set F of allowed non-associated features.

Constraints on the distribution of terminal nodes follow from the feature calculus on LIs developed by [18,30,32]. A function h maps each node $v \in V$ in t to a tuple of feature sequences representing all the non-associated features on nodes dominated by v. If v is a lexical item, it does not dominate any nodes other than itself, and $h(v)$ only contains v's own feature sequence. If v is an interior node, h manipulates the feature sequences of v's children depending on v's label in the following way:

Definition 1 (Merge). *Given a node $v \in V$ with* label$(v) =$ Merge, *and two nodes* $u, w \in V$ *such that* $v \blacktriangleleft u$ *and* $v \blacktriangleleft w$ *and* $h(u) = \langle \texttt{=x}f_2 \cdots f_k, \phi_1, \ldots, \phi_l \rangle$ ($\texttt{x} \in$ **sel**, $f_i \in$ Feat *for* $1 \leq i \leq k$ *(i.e., the sequence* $f_2 \cdots f_k$ *is potentially empty),* $\phi_i \in$ Feat* *for* $0 \leq i \leq l$) *and* $h(w) = \langle \texttt{x}g_2 \cdots g_m, \psi_1, \ldots, \psi_n \rangle$ ($g_i \in$ Feat *for* $1 \leq i \leq m$, $\psi_i \in$ Feat* *for* $0 \leq i \leq n$), *then* $h(v) = merge_1(h(u), h(w))$ *combines* $h(u)$ *and* $h(w)$ *into a single sequence of sequences of features as defined below:*

$$\frac{\langle \texttt{=x}f_2 \cdots f_k, \phi_1, \ldots, \phi_l \rangle \qquad \langle \texttt{x}g_2 \cdots g_m, \psi_1, \ldots, \psi_n \rangle}{\langle f_2 \cdots f_k, \phi_1, \ldots, \phi_l, g_2 \cdots g_m, \psi_1, \ldots, \psi_n \rangle} \; merge_1$$

If the initial features of the initial sequences of both $h(u)$ *and* $h(w)$ *don't match such that they are of the same type* $\texttt{x} \in$ **sel** *and differ in their polarity,* $merge_1(h(u), h(w))$ *is undefined.*

Definition 2 (Move). *Given a node* $v \in V$ *with* label$(v) =$ Move, *and a node* $u \in V$ *such that* $v \blacktriangleleft u$ *and* $h(u) = \langle \texttt{+x}f_2 \cdots f_k, \phi_1, \ldots, \phi_{i-1}, \texttt{-x}g_2 \cdots g_l, \phi_{i+1}, \ldots, \phi_n \rangle$ ($\texttt{x} \in$ **lic**; $f_j, k, g_j, l, \phi_j, n$ *as above) such for all* $j : 1 \leq j \leq n, j \neq i$: ϕ_j's *initial feature is not* $\texttt{-x}$, *then* $h(v) = move_1(h(u))$ *is defined as given below:*

$$\frac{\langle \texttt{+x}f_2 \cdots f_k, \phi_1, \ldots, \phi_{i-1}, \texttt{-x}g_2 \cdots g_l, \phi_{i+1}, \ldots, \phi_n \rangle}{\langle f_2 \cdots f_k, \phi_1, \ldots, \phi_{i-1}, g_2 \cdots g_l, \phi_{i+1}, \ldots \phi_n \rangle} \; move_1$$

If the initial feature of $h(u)$'s *initial sequence is not a licensor feature* $\texttt{+x}$, *and if there is not exactly one sequence of features in* $h(u)$ *whose initial feature is a licensee feature* $\texttt{-x}$ *of the same type,* $move_1(h(u))$ *is not defined.*

These definitions do not make reference to the ordering of phonological features on LIs. I assume that all linear order is established by post-syntactic mapping to derived structures.

The root condition can then be stated as: for r the root node, $h(r) = \langle \mathbf{z}, \varepsilon, \dots, \varepsilon \rangle$ such that $\mathbf{z} \in F$.

Definition 3 (MG). *An MG $G = \langle N, T, S, P \rangle$ is a regular tree grammar with $T = \text{Lex} \cup \{\text{Merge, Move}\}$. The non-terminals N are tuples of sequences of features occurring in Lex, $N \subset (\text{Feat}^+)^+$, namely $N = \{\langle s_0, \dots, s_{|\text{lic}|} \rangle : s_i \in \{\beta : \alpha\beta = \text{features}(l), l \in \text{Lex}\}, 0 \leq i \leq |\text{lic}|\}$, i.e., all $(|\text{lic}| + 1)$-tuples made up of feature sequences that occur as suffix of the feature sequence of some lexical item. The start symbol $S = \langle \mathsf{C}, \varepsilon, \dots, \varepsilon \rangle \in N$, $\mathsf{C} \in F$. The production rules P are defined based on the feature calculus operations merge and move; for $A_i, A_j, A_k \in N$:*

1. $A_i \rightarrow \text{Merge}(A_j, A_k)$ *if $A_i = merge(A_j, A_k)$*
2. $A_i \rightarrow \text{Move}(A_j)$ *if $A_i = move(A_j)$*
3. $A_i \rightarrow l$ *if $A_i = features(l)$, $l \in \text{Lex}$,*

where $X(Y, Z)$ stands for $X \blacktriangleleft Y$ and $X \blacktriangleleft Z$.

We now turn to implementing DCT as further constraints on MGs, which as [16] has shown can be implemented as refined **sel** features iff the constraint is MSO-definable.

3   Dependent-Down Case is Regular

This section shows that dependent-down case can be formalized as MSO-constraint, thereby proving that an MG with Dependent Case Theory falls into the same complexity class as a standard MG: the class of regular tree grammars.

3.1   Defining the MSO-Logic

We define an MSO-logic for derivation trees $t = \langle V, \trianglelefteq, label \rangle$ as generated by an MG G. To do so, we define a model structure $\mathcal{D} = \langle D, F, P^1, P^2 \rangle$ containing a domain D, a set F of functions on D, a set P^1 of first-order predicates, and a set P^2 of second-order predicates whose domain are tuples of unary first-order predicates; a signature $\Sigma = \langle FS, PS^1, PS^2, VS \rangle$ of function symbols FS, first-order predicate symbols PS^1, second-order predicate symbols PS^2, and variable symbols VS; and a signature interpretation Φ, which maps $FS \rightarrow F$, $PS^1 \rightarrow P^1$, and $PS^2 \rightarrow P^2$. Our domain D is the set of nodes V in t. P^1 contains the characteristic functions of $\trianglelefteq$ and $\blacktriangleleft$ such that $\trianglelefteq^\uparrow : (x, y) \mapsto \mathbf{t}$ if $x \trianglelefteq y$, $\mathbf{f}$ else; and $\blacktriangleleft^\uparrow : (x, y) \mapsto \mathbf{t}$ if $x \blacktriangleleft y$, $\mathbf{f}$ else; whose symbols are $\trianglelefteq$ and $\blacktriangleleft$, respectively. Furthermore, for every label σ in the range of $label$, P^1 contains a predicate labeled $Label_\sigma$ which identifies the nodes that have this label:

(9) $\Phi(Label_\sigma) : n \mapsto \mathbf{t}$ if $label(n) = \sigma$, $\mathbf{f}$ else.

Similarly, P^1 contains for each feature f a predicate labeled $Feat_f$ identifying the nodes labeled with lexical items that contain this feature.

(10) $\Phi(Feat_f) : n \mapsto \mathbf{t}$ if $label(n) = l$ for some $l \in Lex$, and $f \in features(l)$;
             $\mathbf{f}$ else.

The predicate labeled Occ holds of a branching node m and a non-branching node n iff m is an occurrence of some feature on the label of n.

(11) $\Phi(Occ) : (m, n) \mapsto \mathbf{t}$ if $label(n) = l$ for some $l \in Lex$, and m is an occurrence of some feature f in l; $\mathbf{f}$ else.

Furthermore, P^1 contains predicates labeled $assigns_{c,i}$ for each inherent case c in the natural language under discussion and for each integer i between 1 and the maximal number of positive features on LIs in the range of $label$.

(12) $\Phi(assigns_{c,i}) : n \mapsto \mathbf{t}$ iff n is a non-branching terminal node labeled with some $l \in Lex$, and l has at least i positive features, and l's i-th positive feature f_i is $\in$ **sel**, and l assigns inherent case c via f_i; $\mathbf{f}$ else.

P^1 and P^2 contain predicates for all possible extensions. For all subsets $A \subseteq D$, P^1 contains a predicate $A^\uparrow$ identifying the nodes in A, and for all subsets $B \subseteq P^1$ of unary first-order predicates, P^2 contains a predicate $B^\uparrow$ identifying the unary first-order predicates in B.

The function labeled $Sliceroot$ takes a node n that is either labeled with a lexical item l or projected by a lexical item l, and returns the highest node that l projects. For **c** l's unique category feature:

(13) $\Phi(Sliceroot) : n \mapsto$ the Merge-node projected by $=\mathbf{f}_k$
                                                    if $label(n) = [\, p :: \ldots \; =\mathbf{f}_k \; \mathbf{c} \ldots \,]$;
                             the Move-node projected by $+\mathbf{f}_k$
                                                    if $label(n) = [\, p :: \ldots \; +\mathbf{f}_k \; \mathbf{c} \ldots \,]$;
                             n if $label(n) = [\, p :: \mathbf{c} \ldots \,]$;
                             undefined   else.

For a full definition see [15, p. 61].

Furthermore, for each integer i between 1 and the maximal number of positive features on LIs in Lex, F contains a function labeled $Proj_i$ that takes a non-branching node n and returns the branching node m projected by n's i-th positive feature:

(14) $\Phi(Proj_i) : n \mapsto$ the Merge/Move-node projected by n's i-th positive feature if n is labeled with an LI that has at least i positive features, **undefined** else.

It follows that for all nodes n, m, $Proj_0(n) = n$, and $Proj_{i+1}(n) = m$ iff $m \blacktriangleleft Proj_i(n)$ and n's label has at least $i + 1$ positive features.

3.2   Defining Shorthand Predicates

Before giving the full MSO-constraint capturing the rules of Dependent Case Theory, we will introduce some shorthand predicates (inspired by "pseudo-features" in [14,16]). These are not predicates in the model structure $\mathcal{D}$, but abbreviations of MSO-formulas that serve no other purpose than to improve readability.

First, we define three index sets stating the different cases for the natural language that we model. For example, in Turkish,

(15)   a. INH $= \{$ABL, DAT, LOC$\}$
       b. DEP $= \{$ACC, DAT$\}$
       c. UNM $= \{$NOM, GEN$\}$

Note that dative (DAT) is both an inherent and a dependent case. This is to demonstrate that this formalism can deal with such ambiguities and with systematic syncretism in general.

We also specify which case is the default case in this language. For Turkish, DEF = NOM.

Second, for each case c we specify a set $compat_c$ of LIs whose exponent is compatible with this case. In the following, we demonstrate the formalism by using one element of each index set. Without loss of generality, we choose ABL for inherent cases, ACC for dependent cases, and NOM for unmarked cases. For Lex the lexicon of G:

(16)   a. $compat_{\text{ABL}} = \{x \in D : label(x) = l, l \in Lex, \text{N} \in features(l),$
                    $exponent(l)$ is ablative$\}$
       b. $compat_{\text{ACC}} = \{x \in D : label(x) = l, l \in Lex, \text{N} \in features(l),$
                    $exponent(l)$ is accusative$\}$
       c. $compat_{\text{NOM}} = \{x \in D : label(x) = l, l \in Lex, \text{N} \in features(l),$
                    $exponent(l)$ is nominative$\}$

If a leaf in t has an exponent that is syncretic between multiple cases, this leaf will appear in multiple sets.[6] For each leaf, we will define a shortcut predicate stating the "actual" (syntactic) case it has, out of all the cases it is morphologically compatible with. We will call these predicates inh_{ABL}, dep_{ACC}, etc. This determination depends on the leaf's context: is it selected for by an LI that requires a certain case? Does it stand in a dependent case licensing configuration? What is its case assignment domain?, and so on. In order to define these "actual" case predicates, we first introduce some auxiliary predicates. These predicates hold of a tuple of nodes iff the formula on the right-hand side of $\Leftrightarrow_{\text{def}}$ evaluates to **t**.

---

[6] For linguistic arguments in support of this approach, see the feature indeterminacy problems discussed in [8–10].

A node x c-commands a node y iff x does not strictly dominate y (this excludes $x \equiv y$), and all nodes z that strictly dominate x also stricly dominate y (this excludes y dominating x):

(17) $ccom(x, y) \Leftrightarrow_{\mathrm{def}} \neg x \trianglelefteq y \wedge (\forall z : z \triangleleft x \rightarrow z \triangleleft y)$

A node x is a phase head iff its category feature is either C or Voice or D. (Other definitions are possible depending on one's theory of phases.)

(18) $phase(x) \Leftrightarrow_{\mathrm{def}} Feat_C(x) \vee Feat_{\mathtt{Voice}}(x) \vee Feat_D(x)$

A node x is in case assignment domain i if there is a phase head y with category feature i that c-commands x and there is no phase head z distinct from y that c-commands x but does not also c-command y.

(19) $domain_i(x) \Leftrightarrow_{\mathrm{def}} (\exists y : phase(y) \wedge Feat_i(y) \wedge ccom(Sliceroot(y), x) \wedge$
$(\forall z : (phase(z) \wedge ccom(Sliceroot(z), x)) \rightarrow$
$(z \equiv y \vee ccom(Sliceroot(z), y))))$

Two nodes x and y are in the same case assignment domain iff all phase heads z that c-command one also c-command the other:

(20) $samedomain(x, y) \Leftrightarrow_{\mathrm{def}} (\forall z : phase(z) \rightarrow$
$(ccom(Sliceroot(z), x) \leftrightarrow ccom(Sliceroot(z), y)))$

We are now ready to define when a leaf has a particular actual inherent case: A nominal leaf x with potential inherent case c has c as its actual case iff x is merged by an LI y that selects for c.

(21)   For k the maximum number of positive features on LIs in the range of label:
$inh_{\mathrm{ABL}}(x) \Leftrightarrow_{\mathrm{def}} compat_{\mathrm{ABL}}^{\uparrow}(x) \wedge$
$(\exists m : m \blacktriangleleft Sliceroot(x) \wedge$
$(\exists n : \bigvee_{1 \leq i \leq k}(Proj_i(n) \equiv m \wedge assigns_{\mathrm{ABL}, i}(n))))$

Since each inh_c for some case c is a pseudonym for a 1-place predicate $D \rightarrow \{\mathbf{t}, \mathbf{f}\} \in P^1$, we can define a set of these predicates, (22). This then provides us with the second-order predicate "$inh^{\uparrow}$".

(22) $inh = \{inh_i : i \in \mathrm{INH}\}$

A potential dependent-down case is an actual case iff the nominal x has the required exponent and does not carry inherent case and is c-commanded by another nominal y in the same case assignment domain that does not have an actual inherent case. The case assignment domain is specified for each case, here C for ACC.

(23) $dep_{\mathrm{ACC}}(x) \Leftrightarrow_{\mathrm{def}} compat_{\mathrm{ACC}}^{\uparrow}(x) \wedge domain_C(y) \wedge (\forall p : inh^{\uparrow}(p) \rightarrow \neg p(x)) \wedge$
$(\exists y : samedomain(x, y) \wedge (\forall p : inh^{\uparrow}(p) \rightarrow \neg p(y)) \wedge$
$ccom(Sliceroot(y), x))$

Note that dep_{ACC} contains the second-order predicate $inh^{\uparrow}$. Strictly speaking, this is not necessary. We could have expressed $(\forall p : inh^{\uparrow}(p) \rightarrow \neg p(x))$ as the conjunction $\neg inh_{\mathrm{ABL}}(x) \wedge \neg inh_{\mathrm{LOC}}(x) \wedge \neg inh_{\mathrm{DAT}}(x)$, iff the natural language we model contains exactly these inherent cases. As we will see shortly, this holds for all second-order shortcut predicates. Thus the present characterization of Dependent Case Theory is in fact first-order definable.[7] However, we will continue the second-order notation for readability purposes.

For a dependent-up case, replace $ccom(Sliceroot(y), x)$ with $ccom(Sliceroot(x), y)$.

We define the set of all shorthand predicates denoting actual dependent case:

(24) $dep = \{dep_i : i \in \mathrm{DEP}\}$

Potential unmarked cases are licensed as actual cases if the structural conditions for dependent cases are not met. DCT uses assignment rules where the assignment of accusative takes precedence over the assignment of nominative. Thus it is excluded that nominative could arise in a structural environment where accusative is attested. However, when employing a licensing view to DCT, we not only need to make sure that accusative is licensed only if there is a c-commander, but also that nominative is licensed only if there is no c-commander. (For domains with a dependent-down case like CP in ergative languages: absolutive is licensed iff it doesn't c-command another eligible NP.[8] For domains with both dependent-up and dependent-down cases: both conditions must hold.) We thus need to treat unmarked cases as Baker's [2] negative dependent cases, although for a different reason. For the formalism, this does not matter: we remain inside MSO-definability.

(25) $unm_{\mathrm{NOM}}(x) \Leftrightarrow_{\mathrm{def}} compat_{\mathrm{NOM}}{}^{\uparrow}(x) \wedge domain_{\mathrm{C}}(x) \wedge$
$(\forall p : (inh^{\uparrow}(p) \vee dep^{\uparrow}(p)) \rightarrow \neg p(x)) \wedge$
$\neg(\exists y : samedomain(y, x) \wedge ccom(Sliceroot(y), x) \wedge$
$(\exists q : (dep^{\uparrow}(q) \vee unm^{\uparrow}(q)) \wedge q(y)))$

(26) $unm = \{unm_i : i \in \mathrm{UNM}\}$

A potential default case becomes an actual case if the nominal's potential inherent, dependent, and unmarked cases are all not its actual case.

(27) $def(x) \Leftrightarrow_{\mathrm{def}} compat_{\mathrm{DEF}}{}^{\uparrow}(x) \wedge$
$(\forall p : (inh^{\uparrow}(p) \vee dep^{\uparrow}(p) \vee unm^{\uparrow}(p)) \rightarrow \neg p(x))$

---

[7] This is a welcome result. MSO can capture many patterns that we do not expect to find in the case distribution in natural language. For instance, MSO-logic can implement modulo-counting, i.e., for a sequence of NPs, the case of the NP depends on its position in the sequence modulo some integer.

[8] This follows the traditional formalization of DCT [23]. More recently, data have shown that ERG–ERG–ABS patterns are unattested [2,24] and realized as ERG–ABS–ABS instead. This paper does not aim to contribute to this issue.

We now have predicates determining for each node n whether n is a nominal LI carrying a certain syntactic case. The predicates are defined in such a way that out of $inh \cap dep \cap unm \cap \{def\}$, only one of these predicates will be t of n.

3.3   Defining the MSO-constraint

We are now ready to formulate a version of the Case Filter that assures that every nominal will have a syntactic case. We state the following MSO-constraint on derivation trees generated by our MG:

(28) $\forall x : Feat_{\mathsf{N}}(x) \rightarrow (\exists p : (inh^{\uparrow}(p) \lor dep^{\uparrow}(p) \lor unm^{\uparrow}(p) \lor \{def\}^{\uparrow}(p)) \land p(x))$

We walk through some scenarios to get an impression of a completeness and soundness proof for (28) with respect to the algorithm in Fig. 1. (28) is complete iff every syntactic tree whose cases have been assigned by Fig. 1 fulfills (28). (28) is sound iff every tree that fulfills (28) will be assigned the same cases by Fig. 1.

The algorithm assigns exactly one case to any NP. It doesn't assign more than one case because it exits the algorithm as soon as a case is assigned. It doesn't assign no case at all because of its catch-all assignment of default case. Similarly, (28) doesn't license more than one syntactic case on any NP due to the definitions of dependent, unmarked, and default cases ensuring that no case lower on the case assignment hierarchy in (1) is licensed as well.

It is easy to see that for intransitives, the algorithm will assign an unmarked case corresponding to the domain the single NP is in, and (28) will only license this unmarked case on the single NP. For regular transitives, the algorithm will assign a dependent case to one of the two NPs, and an unmarked case to the other. (28) will license exactly these cases as well. For transitives that assign an inherent case to one argument α, the algorithm will assign this inherent case to α and unmarked case to the other argument (for languages of the Icelandic-type). (28) will only license this assignment.

Incorrect configurations include dependent case on the single NP argument of an intransitive. The algorithm will not assign this. (28) will not license this. Another incorrect configuration is a regular transitive with both arguments marked with the same type of dependent case (both up or both down). The algorithm will not assign this. (28) will not license this. Finally, the algorithm will not license unmarked case on two NPs standing in a c-command relation. (28) will also not license this, due to the explicit condition on unmarked cases that there must not be a c-commanding/c-commanded NP in the same case assignment domain.

3.4   Conclusions

Graf [16] has shown that any constraint that can be defined in MSO logic can be added to an MG that is strongly equivalent to an MG without this constraint. In this section, we saw that the distribution of morphological case as defined by DCT (Sect. 1.2) can be formalized as MSO constraint. Thus it follows that the rules of DCT, including one-to-many licensing in the assignment of dependent-down cases, is not more complex than other constraints commonly imposed onto MGs.

We now turn to a new question: What is the relationship between the class of constraints that have an MSO-formalization, to the class of constraints captured by Move-dependencies? Are they equivalent, like MSO-constraints and Merge-dependencies, or is one more expressive than the other? Schematically,

(29)   Merge $\Leftrightarrow$ MSO $^?\Leftrightarrow^?$ Move

We take a step towards exploring this question in the next section by attempting to formalize the rules of Dependent Case Theory as Move-dependencies.

4   Dependent-Down Case as Move is Non-regular

Alternatively to specifying the distribution of cases via MSO-statements about the configurations these cases can appear in, we treat Case as a feature on lexical items with a compatible exponent and "check" it against a licensor. This checking can happen at a distance and corresponds to the deletion of "uninterpretable" features under Agree in early Minimalist literature; [6], et seq., i.a. We explore the option of taking cases to be licensee features $-\mathtt{f}$ on the nominal that carries case.

For dependent-down case, we take the c-commanding nominal as licensor. In a clause with a transitive verb, accusative case is assigned to the object in the presence of the c-commanding subject. This is schematized in (30).

(30)

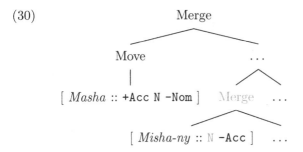

4.1   Generalizing the Movement Type

Our definition of the distribution of Move-nodes and the application of the operation *move* in Definition 2 does not work for (30). Applying $move_1$ to the feature sequence on the child of Move, $\langle$+Acc N -Nom$\rangle$, is undefined. This problem arises because the only kind of movement defined in Sect. 2 was raising movement: the Move node must dominate the LI x carrying the negative feature; thereby the LI y projecting this Move node must have selected for the subtree containing x. However, in (30), this is not the case. The projecting LI *Masha* does not select for anything; rather, it connects to the subtree containing *Misha-ny* by being selected.

This mismatch arises because the movement in (30) is not raising. Graf [15] has shown that MGs can handle other (generalized) movement types besides raising, including lowering and sideward movement, different sizes of the moved constituent (head, phrase, pied-piped phrase), and different linearizations (covert, to the left, to the right). In order to define these movement types, he recognizes that the requirement that a Move-node m can be associated to a feature -f on an LI l only if m dominates l, is a special case of an MSO-definable relation between m and l.

(31) $R^{\text{RAISING}}(m, l)$ iff $m \trianglelefteq l$

Other movement types may be defined in terms of different relations. The movement type DD employed by dependent-down case licensing uses the relation R^{DD} between a Move-node m associated with feature -f on a lexical item l, (32), and the movement type DU for dependent-up case licensing the relation R^{DU} in (33):

(32) $R^{\text{DD}}(m, l)$ iff $ccom(Sliceroot(m), l)$
(33) $R^{\text{DU}}(m, l)$ iff $ccom(Sliceroot(l), m)$

Like with raising, it holds that every -f must have a Move-node as occurrence, and that every Move-node can only be an occurrence of one negative -f feature. Also like raising, we impose a minimality requirement: the Move-node that is an occurrence of a feature -f must be the closest possible Move-node in terms of R^{DD}. Additionally, DD movement is bounded by case assignment domains. Both the Move-node and the LI with -f must be in the same domain, and this domain must be of a particular type.

After [15], we define a constraint on which Move-nodes can be associated to which -Acc features: For a non-branching node l labeled with -Acc and a Move-node m, m is an occurrence of l iff they stand in relation R^{DD} and are in the same case assignment domain which is a CP, and there is no other non-branching node x with either -Acc or +Acc that is closer in terms of R^{DD}:

(34) $\forall m : \forall l : (Label_{\text{Move}}(m) \wedge Feat_{\text{-Acc}}(l)) \rightarrow$
$(Occ(m, l) \leftrightarrow (R^{\text{DD}}(m, l) \wedge domain_C(m) \wedge samedomain(m, l) \wedge$
$(\forall x : (Feat_{\text{-Acc}}(x) \vee Feat_{\text{+Acc}}(x)) \rightarrow$
$(R^{\text{DD}}(x, l) \rightarrow R^{\text{DD}}(x, m)))))$

4.2   Generating Non-regular Tree Languages

Recall the two configurations for dependent-down case assignment discussed in Sect. 1.2. In the 1:n-configuration, a single nominal n licenses case on multiple accusative-marked lexical items. In our system above, this corresponds to n projecting multiple Move-nodes, which each stand in a R^{DD} relation with one of the case-marked LIs.

However, (34) does not define which Move-node is an occurrence of which
-Acc. This ambiguity is illustrated in (35).

(35)

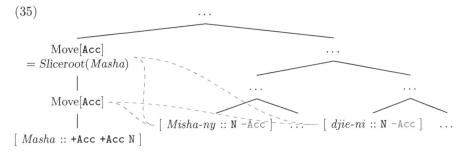

This is not a problem per se as nothing in the grammar depends on which Move-node associates with which LI. I will argue that any derivation will succeed as long as there exists a bijection.

However, this implies that the grammar allows a single licensor to enter into arbitrarily many licensing relationships. In this section, we have silently been assuming that the lexicon contains multiple versions of every non-inherently case-marked nominal: one with +Acc and one without. However, if a nominal can carry arbitrarily many +Acc features, the lexicon will grow from finite to infinite. Since the non-terminals in our tree grammar are made up of feature sequences on lexical items, the grammar will no longer be regular. This constitutes a significant increase in complexity of our formalism, which we would like to avoid in the absence good reason to adopt it. The proof is straight-forward, for instance with the pumping lemma for regular tree languages [7, Sect. 1.2]. Section 5 will propose an amendment to the formalism that allows us to remain in the realm of regularity.

4.3   Conclusions

In this section, we have seen that an MG that formalizes dependent-down case as **lic**-features checked by the c-commanding nominal does not generate regular derivation tree languages. This is due to there being no restriction on the number of instances of dependent-down case that a single nominal can license, and to the requirement in our grammar of a one-to-one relationship between Move-nodes projected by +Acc features and -Acc features (the SMC). Regular languages cannot count to arbitrarily large numbers.

However, in Sect. 3 I showed that in general a tree language with an equivalent distribution of accusative-marked nominals as L *is* regular and expressible by MGs by refining its **sel**-features via MSO-constraints. It follows that a formalization with case as **lic**-features is not equivalent to a formalization with case as **sel**-features.

In the next section, we will explore a possibility to use **lic**-features while not increasing the complexity of the derived tree language beyond regular tree languages. We will achieve this by suspending the SMC for selected features.

5 Selectively Suspending the SMC

The formalization in Sect. 4 increases the complexity of MGs in two ways. First, for dependent-down case licensing in 1:n-configurations, the licensor carries an +Acc feature for every licensing relationship. This results in unboundedly large lexical items and thereby an infinite lexicon. The second problem is the requirement that the number of +Acc features must match the number of Move-nodes projected by these +Acc features, and also the number of −Acc features on the licensees resp. the number of licensees. This requires our tree grammar to count, which is not possible for a regular tree grammar. This stands in clear contrast to the results of Sect. 3.

5.1 Persistent Licensor Features

We can dissociate the number of +Acc features on the licensor from the number of projected Move-nodes by allowing a single +Acc to project multiple Move-nodes. Taking inspiration from Stabler's [31] persistent licensee features, we allow the licensor feature +Acc to be persistent. We add the rule in (37) in addition to the existing $move_1$-rule from Definition 1,

(36)
$$\frac{\langle \text{+Acc}\, f_2 \cdots f_k, \phi_1, \ldots, \phi_{i-1}, \text{−Acc}\, g_2 \cdots g_l, \phi_{i+1}, \ldots \phi_m \rangle}{\langle \text{+Acc}\, f_2 \cdots f_k, \phi_1, \ldots, \phi_{i-1}, g_2 \cdots g_l, \phi_{i+1}, \ldots \phi_m \rangle} \; move_2$$

We also amend the definition of MGs from Definition 3 to contain the production rule in (37):

(37) $A_i \rightarrow \text{Move}(A_j)$ if $A_i = move_2(A_j)$

This will restrict the necessary lexical items to 2 per noun.[9] We have now solved the first of our two problems: the number of +Acc features no longer needs to

---

[9] We could even reduce this to a single lexical item per noun. This is irrelevant for the formalism because the blow-up to the lexicon is constant and thereby negligible, but it does have linguistic significance. Given that the phonological content of a potential licensor virtually never changes with respect to whether it is a licensor or not (J. Bobaljik, p.c.), we would not like our formalism to employ accidental homophony to capture this because doing so would allow us to derive non-homophonous pairs of licensors, thereby overgenerating with respect to the linguistic data. We have two options to capture this systematic syncretism. The first one is to assume a generative lexicon that contains entries of the one type (either licensors or not) and derives entries of the second type. The second option is to introduce a rule that deletes +Acc from a nominal and applies optionally, (i), and to add the scheme for production rules in (ii) to our MG.

(i)
$$\frac{\langle \text{+Acc}\; f_2 \cdots f_i \; \text{N} \; f_j \cdots f_k, \phi_1, \ldots, \phi_n \rangle}{\langle f_2 \cdots f_i \; \text{N} \; f_j \cdots f_k, \phi_1, \ldots, \phi_n \rangle} \; del$$

(ii) $A_i \rightarrow A_j$ if $A_i = del(A_j)$; $A_i, A_j \in N$.

match the number of projected Move-nodes and of `-Acc` features in the derivation. However, the second problem remains: the number of Move-nodes needs to
match the number of `-Acc` features. In Sect. 5.2, I propose a solution to both
problems by keeping the number of `-Acc` features finite.

5.2 Suspending the SMC

The solution lies in abandoning the requirement that each Move-node can only
associate with one `-Acc` feature. If a single Move-node could associate with
unboundedly many `-Acc` features, we would lose the one-to-one requirement
that cannot be expressed by regular tree grammars.

This requirement is known as the SMC and can be paraphrased as "Every
Move-node can be associated with one negative feature." What I propose is
to change this definition to: "Every Move-node that is not projected by `+Acc` is
associated with exactly one negative feature." The association of one Move-node
to arbitrarily many `-Acc` features allows the formalism to not distinguish between
the exact number of `-Acc` features and Move-nodes, but to only differentiate
between two states: "at least one `-Acc` feature needs checking" and "no `-Acc`
feature needs checking". We alter the functions $merge_1$ and $move_1$ defined above
to implement this. `-Acc` features are deleted when they do not contribute to the
distinction between these two states, i.e., when there already is another `-Acc`
awaiting association with a Move-node.

We define the function $merge_2$ that reduces two `-Acc` to one. For $x \in$ **sel**,
$f_{j_1}, g_{j_2}, h_{j_3}, i_{j_4} \in Feat$, $0 \le j_1 \le k$, $0 \le j_2 \le l$, $0 \le j_3 \le m$, $0 \le j_4 \le n$; and
$\Phi, X, \Psi, \Omega \in (Feat^*)^*$:

$$(38) \quad \frac{\langle =xf_2 \cdots f_k, \Phi, -\text{Acc } g_2 \cdots g_l, X \rangle \quad \langle xh_2 \cdots h_m, \Psi, -\text{Acc } i_2 \cdots i_n, \Omega \rangle}{\langle f_2 \cdots f_k, \Phi, -\text{Acc } g_2 \cdots g_l, X, h_2 \cdots h_m, \Psi, i_2 \cdots i_n, \Omega \rangle} \; merge_2$$

We alter the definition of $merge_1$ in Definition 1 on p. 10, repeated as (39),
to apply subordinately to (38), i.e., only if it is not the case that for some
$i \le l, j \le n$, ϕ_i and ψ_j start with `-Acc`:

$$(39) \quad \frac{\langle =xf_2 \cdots f_k, \phi_1, \ldots, \phi_l \rangle \quad \langle xg_2 \cdots g_m, \psi_1, \ldots, \psi_n \rangle}{\langle f_2 \cdots f_k, \phi_1, \ldots, \phi_l, g_2 \cdots g_m, \psi_1, \ldots, \psi_n \rangle} \; merge_1$$

We define the function $move_3$ as in (40) to avoid introduction of a second initial
`-Acc` feature. For $x, f_{j_1}, g_{j_2}, h_{j_3} \in Feat$; $0 \le j_1 \le k$, $0 \le j_2 \le l$, $0 \le j_3 \le m$;
$\Phi, X, \Psi \in (Feat^*)^*$; and `-x -Acc` $g_3 \cdots g_l$ preceding or following `-Acc` $h_2 \cdots h_m$:

$$(40) \quad \frac{\langle +xf_2 \cdots f_k, \Phi, -x \; -\text{Acc } g_3 \cdots g_l, X, -\text{Acc } h_2 \cdots h_m, \Psi \rangle}{\langle f_2 \cdots f_k, \Phi, g_3 \cdots g_l, X, -\text{Acc } h_2 \cdots h_m, \Psi \rangle} \; move_3$$

The rule $move_1$ from Definition 2 on p. 10, repeated in (41), applies only when
it is not the case that $g_2 = $`-Acc` and for some $j, 1 \le j \le n$, the first element in
ϕ_j is `-Acc`:

(41)
$$\frac{\langle +\mathbf{x}f_2\cdots f_k, \phi_1, \ldots, \phi_{i-1}, -\mathbf{x}g_2\cdots g_l, \phi_{i+1}, \ldots, \phi_n\rangle}{\langle f_2\cdots f_k, \phi_1, \ldots, \phi_{i-1}, g_2\cdots g_l, \phi_{i+1}, \ldots \phi_n\rangle} \; move_1$$

A Minimalist Grammar licensing dependent-down case with licensor features without increasing the complexity of its derivation tree language beyond regularity, is then defined as follows:

Definition 4 (MG, licensing dependent-down case). *An MG* $G = \langle N, T, S, P \rangle$ *is a regular tree grammar with* $T = Lex \cup \{\text{Merge}, \text{Move}\}$. *The non-terminals* N *are tuples of sequences of features occurring in Lex,* $N \subset (\text{Feat}^+)^+$, *namely* $N = \{\langle s_0, \ldots, s_{|\mathbf{lic}|}\rangle : s_i \in \{\beta : \alpha\beta = features(l), l \in Lex\}, 0 \leq i \leq |\mathbf{lic}|\}$, *i.e., all* $(|\mathbf{lic}| + 1)$-*tuples made up of feature sequences that occur as suffix of the feature sequence of some lexical item. The start symbol* $S = \langle \mathbf{C}, \varepsilon, \ldots, \varepsilon \rangle \in N$, $\mathbf{C} \in F$. *The production rules* P *are defined based on the feature calculus operations* $merge_1, merge_2, move_1, move_3$ *as defined in (39), (38), (41), and (40), respectively. For* $A_i, A_j, A_k \in N$:

1. $A_i \to \text{Merge}(A_j, A_k)$ *if* $A_i = merge_1(A_j, A_k)$ *or* $A_i = merge_2(A_j, A_k)$
2. $A_i \to \text{Move}(A_j)$ *if* $A_i = move_1(A_j)$ *or* $A_i = move_3(A_j)$
3. $A_i \to l$ *if* $A_i = features(l), l \in Lex$.

The new rules $merge_2$ and $move_3$ do not increase the weak nor the strong generative capacity of MGs. The same string language and phrase structure languages are derivable by MGs with MSO-definable constraints, as demonstrated in Sect. 3. This is so because the present movement is not actual movement: no phonological or semantic features are displaced. It serves merely the feature calculus operations determining the distribution of case morphology. The aim of this paper has been to show that this can be achieved with refined **sel**-features as well as with SMC-liberated **lic**-features.

6   Conclusions

I have shown that dependent-down cases pose a problem for MGs because they allow one-to-many licensing relationships involving one licensor and arbitrarily many licensees. MGs with a one-to-one association between licensor nodes and licensee features (i.e., with the SMC) would need to generate non-regular derivation tree languages in order to ensure that the number of licensor nodes and licensee features match.

However, the high-level rules for dependent-down case assignment are well within the range of regular tree languages. I have shown that this is the case by providing a monadic second-order formula, which is known to be imposable onto an MG by refining its Merge-features [16]. We can thus see an asymmetry between the operations Merge and Move: while these constraints can be expressed as refined Merge-features, they cannot be expressed via Move-features while staying in the same complexity class.

I have argued that it is possible to realize dependent-down case as Move-features if the SMC is suspended. This amounts to adding new rules for *merge* and *move* that will discard a newly activated -Acc feature if there is already an active -Acc in the derivation. In this way, we can formalize a single licensor node being able to license arbitrarily many licensee features.

We have motivation to abandon the SMC for case licensing because it affects neither word order at the PF interface nor scope at the LF interface. However, this does not extend to other instances of one-to-many licensing dependencies in natural languages: quantifier raising and multiple wh-fronting affect LF and PF respectively and so cannot be licensed in this way. Note that Gärtner and Michaelis's [12,13] approach of wh-clustering requires the wh-elements to stand in pairwise c-command relationships. This is parallel to the "daisy-chain" configuration of dependent-down case licensing, and does not raise an issue in either construction. The present paper makes an empirical prediction about multiple wh-fronting: On the assumptions in [12,13] about wh-fronting, for constructions in which not all fronted wh-words stand in a c-command relation to another fronted wh-word (in their respective base positions), the PF-order of moved wh-phrases will be either completely free (arbitrary) or completely fixed (determined by post-syntactic linearization rules), but is crucially not determined in the derivation.

I leave an empirical investigation of this phenomenon, as well as of QR and the attestation of unbounded 1:n dependent-down case assignment for future research.

Acknowledgments. Many thanks to Jonathan Bobaljik, Thomas Graf, Tim Hunter, Stefan Kaufmann, Jos Tellings, and Susi Wurmbrand, as well as to three anonymous reviewers for Formal Grammar, and two anonymous reviewers for the NASSLLI Student Session for helpful feedback and stimulating discussions. All errors are my own.

References

1. Akkuş, F.: On Iranian case & agreement, talk given at the University of Vienna, Austria, 15 December 2017
2. Baker, M.: Case: Its Principles and Its Parameters. Cambridge University Press, Cambridge (2015)
3. Baker, M., Bobaljik, J.: On Inherent and Dependent Theories of Ergative Case. Oxford University Press, Oxford (2017)
4. Baker, M., Vinokurova, N.: Two modalities of case assignment: case in Sakha. Nat. Lang. Linguist. Theory **28**(3), 593–642 (2010)
5. Bobaljik, J., Wurmbrand, S.: Questions with declarative syntax tell us what about selection. In: Gallego, Á., Ott, D. (eds.) 50 Years Later: Reflections on Chomsky's Aspects, MIT Working Papers in Linguistics, vol. 77. MIT Press (2015)
6. Chomsky, N.: The Minimalist Program. MIT Press, Cambridge (1995)
7. Comon, H., Dauchet, M., Gilleron, R., Löding, C., Jacquemard, F., Lugiez, D., Tison, S., Tommasi, M.: Tree Automata Techniques and Applications (2008). http://www.grappa.univ-lille3.fr/tata. Accessed 18 Nov 2008
8. Dalrymple, M., Kaplan, R.M.: Feature indeterminacy and feature resolution. Language **76**(4), 759–798 (2000)
9. Dalrymple, M., King, T.H., Sadler, L.: Indeterminacy by underspecification. J. Linguist. **45**(1), 31–68 (2009)

10. Eisenberg, P.: A note on "Identity of Constituents". Linguist. Inq. **4**(3), 417–420 (1973)
11. Ermolaeva, M.: Morphological agreement in minimalist grammars. In: Foret, A., Muskens, R., Pogodalla, S. (eds.) FG 2017. LNCS, vol. 10686, pp. 20–36. Springer, Heidelberg (2018). https://doi.org/10.1007/978-3-662-56343-4_2
12. Gärtner, H.M., Michaelis, J.: On the treatment of multiple-wh-interrogatives in minimalist grammars. In: Hanneforth, T., Fanselow, G. (eds.) Language and Logos. Studia Grammatica, pp. 339–366. Akademie Verlag, Berlin (2010)
13. Gärtner, H.M., Michaelis, J.: In defense of generalized wh-clustering. In: Baglini, R., Grinsell, T., Keane, X., Singerman, A., Thomas, J. (eds.) Proceedings of the 46th Annual Meeting of the Chicago Linguistic Society: The Main Session, pp. 137–146. The Chicago Linguistic Society, Chicago (2014)
14. Graf, T.: Closure properties of minimalist derivation tree languages. In: Pogodalla and Prost [25], pp. 96–111. https://doi.org/10.1007/978-3-642-22221-4
15. Graf, T.: Movement-generalized minimalist grammars. In: Béchet, D., Dikovsky, A. (eds.) LACL 2012. LNCS, vol. 7351, pp. 58–73. Springer, Heidelberg (2012). https://doi.org/10.1007/978-3-642-31262-5_4
16. Graf, T.: Local and transderivational constraints in syntax and semantics. Ph.D. thesis, UCLA (2013). http://thomasgraf.net/doc/papers/PhDThesis_RollingRelease.pdf
17. Keenan, E., Moss, L.: Mathematical Structures in Languages. University of Chicago Press, Chicago (2015)
18. Kobele, G., Retoré, C., Salvati, S.: An automata-theoretic approach to minimalism. In: Rogers, J., Kepser, S. (eds.) Model-Theoretic Syntax at 10, 13–17 August 2007, Organized as Part of ESSLLI 2007 (2007). https://www.yumpu.com/en/document/view/34852154/
19. Kobele, G.M.: Minimalist tree languages are closed under intersection with recognizable tree languages. In: Pogodalla and Prost [25], pp. 129–144. https://doi.org/10.1007/978-3-642-22221-4
20. Kornfilt, J., Preminger, O.: Nominative as 'no case at all': an argument from raising-to-ACC in Sakha. In: Joseph, A., Predolac, E. (eds.) Proceedings of the 9th Workshop on Altaic Formal Linguistics (WAFL9). MIT Working Papers in Linguistics, vol. 76, pp. 109–120. MIT Press (2015)
21. Levin, T.: Successive-cyclic case assignment: Korean nominative-nominative case-stacking. Nat. Lang. Linguist. Theory **35**, 447–498 (2017). https://doi.org/10.1007/s11049-016-9342-z
22. Maling, J.: Of nominative and accusative: the hierarchical assignment of grammatical case in Finnish. In: Holmberg, A., Nikanne, U. (eds.) Case and Other Functional Categories in Finnish Syntax. Mouton de Gruyter, Berlin (1993)
23. Marantz, A.: Case and licensing. In: Westphal, G., Ao, B., Chae, H.R. (eds.) Proceedings of the Eastern States Conference on Linguistics (ESCOL) 1991, pp. 234–253. Cornell University, CLC Publications, Ithaca (1992)
24. Nie, Y.: Why is there NOM-NOM and ACC-ACC but no ERG-ERG? In: Lamont, A., Tetzloff, K. (eds.) Proceedings of NELS 47. vol. 2, pp. 315–328 (2017)
25. Pogodalla, S., Prost, J.-P. (eds.): LACL 2011. LNCS (LNAI), vol. 6736. Springer, Heidelberg (2011). https://doi.org/10.1007/978-3-642-22221-4
26. Poole, E.: A configurational account of Finnish case. University of Pennsylvania Working Papers in Linguistics 21, Article 26 (2015). https://repository.upenn.edu/pwpl/vol21/iss1/26

27. Poole, E.: The locality of dependent case (2016). http://ethanpoole.com/handouts/2016/poole-dependent-case-locality.pdf. Presented at GLOW 39 (April 5) and WCCFL 34 (April 30, no proceedings paper)

28. Preminger, O.: Agreement and its Failures, Linguistic Inquiry Monographs, vol. 68. MIT Press, Cambridge (2014)

29. Salvati, S.: Minimalist grammars in the light of logic. In: Pogodalla, S., Quatrini, M., Retoré, C. (eds.) Logic and Grammar. LNCS (LNAI), vol. 6700, pp. 81–117. Springer, Heidelberg (2011). https://doi.org/10.1007/978-3-642-21490-5_5

30. Stabler, E.: Derivational minimalism. In: Retoré, C. (ed.) LACL 1996. LNCS, vol. 1328, pp. 68–95. Springer, Heidelberg (1997). https://doi.org/10.1007/BFb0052152

31. Stabler, E.: Computational perspectives on minimalism. In: Boeckx, C. (ed.) The Oxford Handbook of Linguistic Minimalism, chap. 27, pp. 617–642. Oxford University Press, Oxford (2011). https://doi.org/10.1093/oxfordhb/9780199549368.001.0001

32. Stabler, E., Keenan, E.: Structural similarity within and among languages. Theor. Comput. Sci. **293**(2), 345–363 (2003)

33. Vinokurova, N.: Lexical categories and argument structure: a study with reference to Sakha. Ph.D. thesis, University of Utrecht (2005)

34. Yip, M., Maling, J., Jackendoff, R.: Case in tiers. Language **63**(2), 217–250 (1987)

A Challenge for Tier-Based Strict Locality from Uyghur Backness Harmony

Connor Mayer[(✉)] and Travis Major

UCLA, Los Angeles, CA 90095, USA
connormayer@ucla.edu

Abstract. In this paper we describe the process of backness harmony in Uyghur, where suffix forms are determined first from the backness of certain vowels in the stem, or, if no such vowels are present, from the backness of dorsals in the stem. We show that this pattern cannot be captured by a tier-based strictly local (TSL) language. This is problematic for the weak subregular hypothesis, which claims that all segmental phonological stringsets are TSL languages. Next, we consider an alternative phonological analysis that is compatible with a TSL representation, but empirically unsupported. Finally, we consider the possibility that Uyghur backness harmony might be a lexicalized pattern, and find some suggestive evidence in support of this. This alternative appears to be the most likely way in which Uyghur backness harmony might, in principle, turn out to be compatible with the hypothesis that TSL languages provide an upper bound on phonological learnability.

Keywords: Uyghur · Vowel harmony · Backness harmony
Subregular hierarchy · Subregular hypothesis · Formal complexity
Phonology · TSL

1 Introduction

Researchers in computational linguistics propose that insights from theories of computation can guide how we study linguistic systems and what predictions we make about the structures of natural language (e.g. [21]). Hypothesizing that some aspect of language is bound by a particular computational structure has the potential to capture the wide variety of patterns seen across languages, while simultaneously constraining the types of patterns we should expect.

It is commonly accepted that phonological processes are regular: that is, they can be computed by regular grammars/automata (e.g. [28,29]). This applies to both phonological stringsets (the properties of surface strings that may be characterized by phonotactic and markedness constraints) and phonological maps (the relationship between underlying and surface forms).

A stronger claim is that all phonological stringsets are tier-based strictly local (TSL) languages, which are subregular. That is, valid stringsets can be expressed as prohibitions on substrings, but these substrings may belong to "tiers" which

© Springer-Verlag GmbH Germany, part of Springer Nature 2018
A. Foret et al. (Eds.): FG 2018, LNCS 10950, pp. 62–83, 2018.
https://doi.org/10.1007/978-3-662-57784-4_4

contain only some subset of the segments in a language [23]. This is referred to as the *weak subregular hypothesis* [21].[1]

There are several existing counterexamples against TSL as an upper bound for phonological complexity. Some suprasegmental patterns have been identified as being outside of TSL, such as culminative quantity-sensitive stress rules [4] and circumambient patterns like unbounded tone plateauing [25]. A handful of segmental patterns that cannot be generated by single TSL grammars are described by McMullin [34]. McMullin claims that some of these exceptions can be captured using the intersection of multiple TSL grammars. However, the more complex patterns require a more powerful system, such as using an Optimality Theory account with constraints based on violations of TSL grammars. de Santo and Graf have formalized the intersection of multiple TSL languages as multi-tier strictly local (MTSL) languages, and proposed an extension of TSL, structure-sensitive TSL (SS-TSL), that allows the more problematic patterns described by McMullin to be captured [13]. Graf has also defined an extension of strictly piecewise (SP) grammars, interval-based SP (IBSP) grammars, which introduces domain restrictions to SP grammars and allows the problematic suprasegmental patterns described above to be captured [18].

This paper provides new data on a phonological process that is beyond the capacity of TSL grammars, providing a counterexample to the weak subregular hypothesis: backness harmony in Uyghur. This pattern is of interest for several reasons. First, it is a *segmental* process that cannot be generated by TSL grammars. These patterns are less common than suprasegmental patterns [25]. Furthermore, this pattern is significantly more complex than any segmental pattern previously discussed: Uyghur backness harmony cannot be captured by *any* of the classes that have been investigated in subregular phonology, with the exception of the overly powerful star-free languages. This makes it a particularly difficult case for anything but the weakest versions of the subregular hypothesis.

The paper is organized as follows. Section 2 will give a brief description of the properties of TSL languages. Section 3 will outline the characterization of backness harmony in Uyghur presented in the literature, and Sect. 4 will show that this pattern cannot be generated by grammars in any of the classes that have been previously investigated in subregular phonology. We briefly sketch an enhancement similar to SS-TSL that is able to capture this pattern, but leave its elaboration and implications for future research.

Given this data, it is either the case that previously considered subregular languages are insufficient to capture all phonological stringsets, or that another analysis for this phenomenon must be adopted. In Sect. 5 we describe an alternative characterization of Uyghur backness harmony that is compatible with a TSL analysis. We then present original experimental data suggesting that this analysis lacks empirical support. With this result in mind, Sect. 6 shows that Uyghur

---

[1] The *strong subregular hypothesis* claims that phonological stringsets are either strictly local (SL) or strictly piecewise (SP) languages [20]. Some autosegmental processes like stress have been claimed to be fully regular (e.g. [17]), though there is debate on whether alternative analyses are possible [20].

backness harmony exhibits many of the characteristics that are frequently taken as evidence of a lexicalized pattern. We elaborate on the implications of this in Sect. 7. If the description of Uyghur backness harmony in the literature is correct, then the weak subregular hypothesis is false. If this conclusion is to be avoided, then the most likely alternative to this description appears to be that the pattern is in fact lexicalized.

2   Tier-Based Strictly Local Languages

Tier-based strictly local languages fall within the subregular hierarchy, shown in Fig. 1. Researchers have identified a variety of subregular language classes, and established their mathematical properties and the relationships between them (e.g. [14, 38, 44, 45]). We will not go into detail here about the subregular hierarchy in general, but many excellent descriptions and applications can be found elsewhere (e.g. [21, 41, 42, 48, 49]).

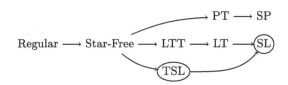

Fig. 1. The subregular hierarchy. Classes we discuss in this section are circled.

The class of TSL languages properly contains the class of strictly local (SL) languages and is properly contained within the class of star-free languages. It is incomparable with other subregular classes [23]. We first describe the properties of SL languages, and then examine how TSL languages expand on these.

Σ represents an alphabet. We will use the alphabet $\Sigma = \{a, b, c\}$ throughout the examples in this section. The symbols $\rtimes$ and $\ltimes$ are initial and final markers respectively, which are not in Σ. We will occasionally omit these for readability. The k-factors of a string $w \in \{\rtimes\} \cdot \Sigma^* \cdot \{\ltimes\}$ are defined as all substrings of w that are of length k, where $A \cdot B = \{ab \mid a \in A, b \in B\}$. A string u is a substring of a string w if $w = xuy$ for some strings $x, y \in \Sigma^*$. We can define a function $F_k(w)$ that returns the set of k-factors of w:

$$F_k(w) = \{u \mid u \text{ is a } k\text{-factor of } w\} \tag{1}$$

For example, $F_2(\rtimes ababac\ltimes) = \{\rtimes a, ab, ba, ac, c\ltimes\}$.

A strictly k-local grammar consists of a finite set of k-factors taken from $(\{\rtimes, \ltimes\} \cup \Sigma)^k$, which describe illicit substrings.[2] A string $w \in \{\rtimes\} \cdot \Sigma^* \cdot \{\ltimes\}$ is well formed with respect to a k-SL grammar G iff $F_k(w) \cap G = \emptyset$, i.e. if it

---

[2] These can equivalently be formulated as licit substrings.

contains no illicit substrings. A language L is SL iff there is some $k \in \mathbb{N}$ such that L can be generated by a strictly k-local grammar.

For example, suppose we want to define a language that prohibits all strings in which b is immediately followed by c. We could define a strictly 2-local grammar $G = \{bc\}$, which rules out strings such as $w_1 = ababca$, because $F_2(w_1) \cap G = \{bc\}$, but permits strings such as $w_2 = abacba$, because $F_2(w_2) \cap G = \emptyset$.

TSL grammars differ from SL grammars in that they are defined over a tier $T \subseteq \Sigma$ [23]. Only the segments on this tier are considered when checking for illicit k-factors. Formally, the representation of a string on a tier is generated by means of an erasing, or projection, function E_T, which removes symbols from the string that are not in T:

$$E_T(\sigma_1 \cdots \sigma_n) = u_1 \cdots u_n \qquad (2)$$

where $u_i = \sigma_i$ iff $\sigma_i \in T$ and $u_i = \lambda$ (the empty string) otherwise. A k-TSL grammar consists of a set of k-factors taken from $(\{\rtimes, \ltimes\} \cup T)^k$. A string $w \in \{\rtimes\} \cdot \Sigma^* \cdot \{\ltimes\}$ is well formed with regard to a TSL grammar G iff $F_k(E_T(w)) \cap G = \emptyset$, i.e. if it contains no illicit substrings when projected on T. A language L is TSL iff it can be generated by a strictly k-local grammar over T for some $k \in \mathbb{N}$.

Suppose we want to define a language where words cannot contain both b and c. SL grammars are unable to capture this. We may define a strictly k-local grammar G that contains the k-factor $ba^{k-2}c$, where a^{k-2} represents the symbol a repeated $k - 2$ times. This factor will rule out words like $aba^{k-2}ca$, but not words like $aba^{k-1}ca$, because the window of length k over which the k-factors operate is too small to see both the b and the c. Increasing k will not help, since it is always possible to increase the number of intervening a's. This is the result of a general property of SL languages [42]:

Theorem 1 (Suffix substitution closure). *A language L is SL iff there is some $k \in \mathbb{N}$ such that if there is a string x of length $k - 1$ and strings u_1, t_1, u_2, and t_2, such that $u_1xt_1 \in L$ and $u_2xt_2 \in L$ then $u_1xt_2 \in L$.*

In contrast, it is trivial for a TSL grammar to capture this by letting $T = \{b, c\}$ and $G = \{bc, cb\}$. Under this formulation, the number of intervening a's is irrelevant, because they are excluded from T.

3   Uyghur Backness Harmony

Uyghur is a southeastern Turkic language with SOV word order. It has roughly 10 million speakers in the Xinjiang Uyghur Autonomous Region in the People's Republic of China and neighboring regions such as Kazakhstan and Kyrgyzstan. It has a rich system of vowel and consonant harmony along several dimensions. We focus here on backness harmony, which requires suffix forms to agree in backness with vowels and certain consonants within a stem.

The Uyghur vowels are shown in Table 1. The vowels behave as front or back as specified in the table, with the exception of /i/ and /e/, which are transparent

Table 1. The Uyghur vowel system. Harmonizing vowels are in bold.

	Front		Back	
	Unrounded	Round	Unrounded	Round
High	i	**y**		**u**
Mid	e	**ø**		**o**
Low	æ		**a**	

Table 2. The harmonizing Uyghur dorsal consonants

	Front	Back
Voiceless	k	q
Voiced	g	ʁ

to harmony processes [30,47].[3] We refer to /i/ and /e/ as *transparent* vowels, and the remainder of the vowels as *harmonizing* vowels. A subset of the dorsal consonants, shown in Table 2, also participate in backness harmony, with velars patterning with front vowels, and uvulars patterning with back vowels.[4]

Native Uyghur stems tend to be harmonious with respect to backness. This is not an absolute requirement for stems, however, and disharmonious stems are especially common in loanwords (e.g. /pæmidur/ 'tomato'). Such stems play a particularly interesting role in harmony processes.

The segments of a large class of Uyghur suffixes are underlyingly unspecified for backness. These suffixes take on the back feature of the stems they attach to. We will use the locative case marker /-DA/ as a representative example throughout the paper, but similar patterns occur with many other suffixes.

The examples in Table 3 provide a representative characterization of the pattern. Each example has a corresponding description of the particular type of harmony it illustrates. We refer back to these examples throughout the paper.

The voicing alternation of the initial segment in the suffix is not important, but note crucially that the vowel changes from front to back depending on the stem. The process for determining the backness value of the suffix is as follows:

1. Match the backness of the final harmonizing vowel in the stem. In (4) the stem is treated as a back stem because the final harmonizing vowel /o/ is back, and in (3) the stem is treated as a front stem even though it contains both front and back vowels because the final harmonizing vowel /æ/ is front.
2. If the stem has no harmonizing vowels, find the final dorsal consonant in the stem and match its backness. Note that in (5), the stem is treated as front even though it has only transparent vowels because the stem contains /g/, while in (6) the stem is treated as back because of its uvulars.

---

[3] Note that these vowels are the only ones in the system that have no counterparts differing only in backness. Because /e/ only occurs in loanwords and as the result of certain phonological processes, we focus primarily on /i/ throughout the paper.

[4] The velar sounds /x/ and /ŋ/ do not harmonize.

Table 3. Examples of Uyghur backness harmony. The alternating suffix is indicated in bold, and the harmony triggers are underlined.

Form	Gloss	Harmony type	
aʁinæ-**dæ** friend-LOC	"on the friend"	Closest front vowel	(3)
qoichi-**da** shepherd-LOC	"on the shepherd"	Closest back vowel	(4)
gezit-**tæ** newspaper-LOC	"on the newspaper"	Closest front dorsal	(5)
qirʁiz-**da** Kyrgyz-LOC	"on the Kyrgyz"	Closest back dorsal	(6)
rak-**ta** shrimp-LOC	"on the shrimp"	Closest back vowel across front dorsal	(7)
mæʃq-**tæ** exercise-LOC	"on the exercise"	Closest front vowel across back dorsal	(8)

Table 4. Examples of stems with arbitrary backness specification. The alternating suffix is indicated in bold.

Form	Gloss	Harmony type	
it-**ta** dog-LOC	"on the dog"	No harmonizers, arbitrarily back	(9)
biz-**dæ** we-LOC	"on us"	No harmonizers, arbitrarily front	(10)

Harmonizing vowels always take precedence over harmonizing dorsals, as (7) and (8) show. In these examples the harmonizing vowel determines the form of the suffix, even though a dorsal with the opposite backness intervenes. The process of only falling back on consonants to determine stem backness when insufficient information from vowels is available is the cause of the difficulties for TSL, as we will see in the next section.

Words with no harmonizing vowels or dorsals are arbitrarily specified for backness.[5] This is shown in Table 4. Such stems are theoretically problematic, but we will set them aside for now and return to them in Sect. 6.

4 The Formal Complexity of Uyghur Backness Harmony

In this section we focus on the pattern involving harmonizing vowels and dorsals described above. Because the actual segmental content is not of crucial

---

[5] There is a statistical tendency for such stems to be treated as back.

importance, we use a more abstract notation to simplify the specification of the grammars. V_f and V_b refer to the sets of front and back harmonizing stem vowels.

$$V_f = \{y, \emptyset, æ\} \tag{11}$$

$$V_b = \{u, o, a\} \tag{12}$$

C_f and C_b refer to the sets of front and back harmonizing dorsal stem consonants.

$$C_f = \{k, g\} \tag{13}$$

$$C_b = \{q, ʁ\} \tag{14}$$

We use the symbols S_f and S_b to refer to the sets of front and back suffix forms.

These symbols comprise an alphabet $\Sigma_h = \{V_f, V_b, C_f, C_b, S_f, S_b\}$. We define a homomorphic mapping function $h : \Sigma^* \mapsto \Sigma_h^*$ that converts strings from the full Uyghur alphabet to the notation described above (i.e. it maps stem symbols individually according to the definitions in (11) to (14), entire suffixes to S_f or S_b, and all other sounds to the empty string λ).

Uyghur backness harmony can be characterized succinctly with the following regular expression, which picks out licit strings. The class of regular languages is closed under homomorphism.

$$(\Sigma_h^* V_f \overline{V_b}^* S_f) | (\Sigma_h^* V_b \overline{V_f}^* S_b) | (\overline{(V_f|V_b)}^* C_f C_f^* S_f) | (\overline{(V_f|V_b)}^* C_b C_b^* S_b) \tag{15}$$

Thus it is clear that this pattern is at most regular.

4.1 Challenges for TSL

In this section we will show that Uyghur backness harmony as described in (15) cannot be captured using TSL languages, but first we must comment on our notation. Although the set of regular languages is closed under relabeling, the set of SL (and hence TSL) languages is not. For example, the language $(ab)^*$ is SL, but its image under the relabeling $\{a \mapsto c, b \mapsto c\}$, $(cc)^*$, is not SL. To avoid an increase in generative capacity, we apply this relabeling to the *grammar* rather than the language. In other words, the relabeling is applied to the k-factors defined in the grammar, and the resulting grammar filters out candidate strings in the image of that relabeling. This provably results in no increase in generative capacity so long as the mapping is many-to-one, as it is here [1].

To deal with the vowel component, we can define a grammar over the tier

$$T_v = V_f \cup V_b \cup S_f \cup S_b \tag{16}$$

containing the following illicit 2-factors:

$$V_f S_b \tag{17}$$

$$V_b S_f \tag{18}$$

These factors rule out forms like *[aʁinæ-da] (cf. (3)) and *[qoichi-dæ] (cf. (4)). Harmony with dorsals can be captured by defining a grammar over the tier

$$T_c = C_f \cup C_b \cup S_f \cup S_b \tag{19}$$

containing the following illicit 2-factors:

$$C_f S_b \tag{20}$$

$$C_b S_f \tag{21}$$

These factors rule out forms like *[gezit-ta] (cf. (5)) and *[qirʁiz-dæ] (cf. (6)). Thus 2-TSL grammars can capture the vowel and consonant patterns in isolation.

The difficulty arises when combining these two patterns into a single TSL grammar. Because harmonizing dorsal and vowel information must be considered simultaneously, we must define a grammar over a tier that contains both the relevant dorsals and vowels:

$$T = T_v \cup T_c \tag{22}$$

The grammar over T must be able to look to the beginning of the string to check for the presence of harmonizing vowels. We extend T to contain ⋊ and use $C = C_f \cup C_b$ for the sake of brevity. Suppose we define k-factors over T for some fixed k of the following form:

$$V_b C^{k-2} S_f \tag{23}$$

$$V_f C^{k-2} S_b \tag{24}$$

$$\rtimes C^{k-3} C_b S_f \tag{25}$$

$$\rtimes C^{k-3} C_f S_b \tag{26}$$

(23) and (24) try to capture harmony with vowels, while (25) and (26) try to capture cases with only harmonizing dorsals. These cannot work for all possible forms. Consider the following word on T:

$$w = V_b C_f^{k-1} S_f \tag{27}$$

This word, which has a mismatch between the final vowel and suffix form, will be erroneously included because the number of post-vowel dorsal consonants is too large for the k-factors to see both the vowel and suffix form at the same time. Checking for the absence of harmonizing vowels is bounded by k under suffix substitution closure, and therefore a TSL grammar over a tier containing both harmonizing dorsals and vowels cannot capture this pattern for arbitrary values of k, placing it outside of TSL.

Another possibility for capturing this pattern is to use the intersection of the TSL languages on T_v and T_c defined in (16)–(18) and (19)–(21) respectively. The class of TSL languages is not closed under intersection, and the resulting language falls in the class of multi-tier strictly local languages (MTSL), which properly contains the class of TSL languages [13]. Even this more powerful formalism cannot capture this pattern. The difficulty arises from the fact that violations on T_c should be ignored unless neither V_f nor V_b appear in T_v. Consider again examples (7) and (8), which we repeat in Table 5 along with their tier-based representations.

Table 5. Examples of Uyghur backness harmony over intervening, conflicting dorsals. The alternating suffix is indicated in bold, and the harmony triggers are underlined.

Form	Gloss	T_v	T_c	Harmony type	
ra̱k-**ta** shrimp-LOC	"on the shrimp"	$V_b S_b$	$C_f S_b$	Closest back vowel across front dorsal	(28)
mæ̱ʃq-**tæ** exercise-LOC	"on the exercise"	$V_f S_f$	$C_b S_f$	Closest front vowel across back dorsal	(29)

Violations on one tier cannot be overlooked given the contents of another, so this grammar rules both (28) and (29) as illicit because they are ill-formed on T_c. It would rule them ill-formed on T_v if suffixes of the opposite backness were used. Thus Uyghur backness harmony is also not MTSL.

4.2   Challenges for Other Subregular Languages

The previous section showed that neither TSL nor MTSL grammars can capture the pattern in (15). We focused on these classes because they have received the most consideration as possible subregular upper bounds for phonotactic complexity. In this section we will sketch the arguments that the other subregular classes of languages that have been applied to phonology, including more expressive extensions of TSL, do not contain this pattern. We do not provide formal definitions of these languages here, but refer the reader to previous work.

Uyghur Backness Harmony is Not SS-TSL or SS-MTSL. Structure sensitive tier-based strictly local (SS-TSL) languages generalize the tier-projection process used in TSL [13]. TSL uses a 1-Input Strictly Local (1-ISL) projection, meaning that the projection function considers each segment in isolation (i.e. whether that segment is a member of T) [10]. SS-TSL generalizes this projection to a k-ISL projection, which means the projection function can consider a window of size k around the target segment. For example, we may define a SS-TSL grammar that will project a segment a to a tier only when it is immediately followed by segment b, but not otherwise. Structure sensitive multi-tier strictly local languages (SS-MTSL) are the intersection of multiple SS-TSL languages.

Intuitively, one might try to capture the Uyghur pattern by projecting harmonizing dorsals only when they are not preceded by a harmonizing vowel. It is easy to show using the suffix substitution closure property discussed at the end of Sect. 2 that cannot work for all forms. Assume a 2-SS-TSL grammar that includes the illicit 2-factor $C_b S_f$. Assume also that the projection function is

k-ISL for some k, with the target segment falling into the final slot in the window (i.e. the context we consider is the target segment plus the preceding $k-1$ segments). The string $V_f C_b^k S_f$ will be excluded from this language even though it is a valid Uyghur word because the last of the k occurrences of C_b will be projected onto the tier. SS-MTSL fails for the same reason.

Uyghur Backness Harmony is Not PT or SP. Piecewise testable (PT) grammars are an extension of strictly piecewise (SP) grammars. SP grammars are similar to SL grammars but prohibit *subsequences* (i.e. precedence relations between segments) rather than substrings [45]. PT languages are the closure of SP languages under the Boolean operators $\wedge$ and $\neg$ [41]. Informally, these languages extend SP with the ability to *require* some subsequence be present in a string.

Even the basic vowel harmony pattern cannot be captured by a PT language. The intuition behind this is that the backness of suffixes is determined by the immediately preceding harmonizing vowel, but PT languages cannot precisely capture the order in which vowels occur. For example, $V_f V_b S_b$ and $V_b V_f S_b$ both contain the subsequences $V_f S_b$ and $V_b S_b$, but the first is a legal word while the second is not. We can show this more formally using the following theorem [41]:

Theorem 2 (k-Subsequence Invariance). *A language L is Piecewise Testable iff there is some $k \in \mathbb{N}$ such that for all strings x and y, if x and y contain the same set of subsequences of length k or less, then either $x \in L$ and $y \in L$ or $x \notin L$ and $y \notin L$.*

Consider the following pair of words for some $k \in \mathbb{N}$:

$$w_1 = (V_f V_b)^k S_b \tag{30}$$

$$w_2 = (V_b V_f)^k S_b \tag{31}$$

These words contain the same subsequences of length k or less,[6] but w_1 is a valid word while w_2 is not. Thus even the simplest subcase of Uyghur backness harmony is not PT, and since PT properly contains SP, it is also not SP.

Uyghur Backness Harmony is Not LTT or LT. Locally threshold testable (LTT) grammars are an extension of locally testable (LT) grammars. LT languages are the closure of the SL languages under the Boolean operators $\wedge$ and $\neg$ [41,42]. Informally, these languages extend SL with the ability to *require* some element be present in a string. LTT languages are the closure of LT languages under the first order logic operators $\forall$ and $\exists$, which quantify over position indices [41,42]. Indices can be compared for equality and successorship. Informally, this

---

[6] This can be shown by induction: both words contain the same subsequences when $k = 1$, and the subsequences added with each increase in k will be the k-subsequences generated by prepending V_f or V_b to all subsequences of length $k-1$.

extension allows LTT languages to count the number of occurrences of each
k-factor up to a certain threshold.

It is simple to show that Uyghur backness harmony as described in (15) is
outside of LTT by appealing to the following theorem [41]:

Theorem 3 (Local Threshold Test Invariance). *A language L is Locally
Threshold Testable iff there is some $k \in \mathbb{N}$ and some threshold $t \in \mathbb{N}$ such that,
for all strings x and y, if for any k-factor w, x and y contain the same number
of occurrences of w or have at least t occurrences, then either $x \in L$ and $y \in L$
or $x \notin L$ and $y \notin L$.*

Consider the following two words for some $k \in \mathbb{N}$:

$$w_1 = (C_f)^{k-1} V_b (C_f)^{k-1} V_f (C_f)^{k-1} S_f \tag{32}$$

$$w_2 = (C_f)^{k-1} V_f (C_f)^{k-1} V_b (C_f)^{k-1} S_f \tag{33}$$

Both have the same number of occurrences of every k-factor, but w_1 is a valid
Uyghur word while w_2 is not. Therefore Uyghur backness harmony is not LTT,
and since LTT properly contains LT, it is also not LT.

Uyghur Backness Harmony is Not IBSP. Interval-based strictly piecewise
(IBSP) grammars are an extension of SP grammars that allow k-subsequences
to be defined over a particular interval, such as a word or a prosodic phrase
[18]. The set of IBSP languages properly contains both TSL and SP languages,
and is properly contained by the star-free languages. Uyghur backness harmony
is a word-level process, and an IBSP grammar that is defined over words will
encounter the same issues as the PT and SP grammars described above. We can
think of no interval below the word that is able to avoid these problems, and so
we conjecture that Uyghur backness harmony is not IBSP.

4.3   A Formal Lower Bound for Uyghur Backness Harmony

The pattern in (15) can be captured by the non-counting (NC) or star-free languages, which are the most expressive subregular languages [35]. NC languages
allow the use of the first order logic operators $\exists$ and $\forall$, which quantify over
position indices in the string. Indices can be compared for equality, using the $\approx$
operator, and precedence, using the $<$ operator. Predicates over indices $P(x)$ are
true if the symbol at index x is P. All of the language classes described above
are properly contained by the class of NC languages.

The following expressions define a NC grammar that captures licit forms
under Uyghur backness harmony.

$$\forall x[S_b(x) \Rightarrow \forall y[V_f(y) \Rightarrow \exists z[V_b(z) \land y < z < x]]] \tag{34}$$

$$\forall x[S_f(x) \Rightarrow \forall y[V_b(y) \Rightarrow \exists z[V_f(z) \land y < z < x]]] \tag{35}$$

$$\forall x[S_b(x) \land \neg \exists y[V_f(y) \lor V_b(y)] \Rightarrow \forall z[C_f(z) \Rightarrow \exists w[C_b(w) \land z < w < x]]] \tag{36}$$

$$\forall x[S_f(x) \land \neg \exists y[V_f(y) \lor V_b(y)] \Rightarrow \forall z[C_b(z) \Rightarrow \exists w[C_f(w) \land z < w < x]]] \tag{37}$$

The first two expressions require suffixes to match the backness of the final harmonizing stem vowel. The latter two require suffixes to match the backness of the final harmonizing stem dorsal if there are no harmonizing stem vowels.

Although further weakening the weak subregular hypothesis to include NC languages captures the data presented above, this is not a desirable result from the perspective of learnability. It has been shown that TSL languages are efficiently learnable in polynomial time from polynomial data [22, 26, 27], while NC languages are not [16]. This makes theories of phonological learning somewhat more problematic. Below we briefly sketch a proposal for a new subregular class that is less powerful than the NC class but still sufficient to capture this pattern.

4.4   OSS-TSL

A generalization of TSL, which we tentatively call output structure-sensitive tier-based strictly local (OSS-TSL), can capture Uyghur backness harmony.[7] SS-TSL generalizes the projection function of TSL from a 1-ISL map to a k-ISL map. The class of 1-ISL maps is identical to the class of 1-output strictly local (1-OSL) maps, meaning the TSL projection function could be equally characterized as a 1-OSL function [10]. We could thus generalize the projection mechanism to be k-OSL. This would allow the tier projection function to consider material already on the tier when choosing whether to project a segment from the input.

The pattern in (15) requires a 2-OSL projection function that behaves as follows: V_f, V_b, S_f, and S_b are always projected, while C_f and C_b are only projected if the previous symbol is not V_f or V_b. In other words, we stop adding dorsals to the tier as soon as we encounter a harmonizing vowel. 2-factors defined over this tier would simply check for backness mismatches between the suffix and the preceding symbol.

It is beyond the scope of this paper to consider the formal properties of OSS-TSL grammars and how widely applicable they will be in describing natural language phonology. We intend to investigate this in future research.

5   An Analysis Without Transparent Vowels

Given the uncommonness of segmental patterns that are as problematic for the weak subregular hypothesis as Uyghur backness harmony, it is worth investigating whether the characterization of the pattern presented above is correct. The issues this pattern poses for TSL representations hinge on backness being determined first from vowels, and then from consonants if the vowels prove insufficient. A possible alternative analysis that is compatible with a TSL representation is that Uyghur in fact has no transparent vowels. Rather, there are two different surface versions of /i/ and /e/ which are not reflected in the orthography or in past descriptions of the phonology, one of which is front and one of which is

---

[7] We also suggest that SS-TSL might be relabeled as input structure-sensitive tier-based strictly local (ISS-TSL).

back. We represent the back variants as /ɨ/ and /ə/ respectively, and refer to the pairs as /I/ and /E/ when backness is not important. Such an analysis makes the 2-factors defined over T_v in the previous section sufficient to capture Uyghur backness harmony: V_f now includes /i/ and /e/, and V_b includes /ɨ/ and /ə/, so no reference to consonants is necessary. This account is supported by historical evidence: Uyghur once had a distinction between the front and back unrounded vowels /i/-/ɯ/ and /e/-/ɤ/ as in Turkish, but these vowels have since collapsed into /i/ and /e/ [30].

Under this formulation, /I/ and /E/ are underlying specified as [± back]. This allows us to tidily capture forms like (9) and (10), which no longer need to be arbitrarily specified as front or back, but now select their suffix based on the quality of the final vowel, as below:

$$\text{ɨt-}\textbf{ta} \qquad \text{"on the dog"} \tag{38}$$
$$\text{dog-LOC}$$

$$\text{biz-}\textbf{dæ} \qquad \text{"on us"} \tag{39}$$
$$\text{we-LOC}$$

The generalization that suffixes tend to match the backness of the final harmonizing vowel in the stem, or, if these are lacking, the final dorsal, can be captured by cooccurrence restrictions: /I/ and /E/ must agree in backness with the nearest harmonizing vowel or dorsal, which gives the appearance of suffixes harmonizing with consonants. Thus we can reanalyze (4), (5), and (6) as below.

$$\text{qoɨchɨ-}\textbf{da} \qquad \text{"on the shepherd"} \tag{40}$$
$$\text{shepherd-LOC}$$

$$\text{gezit-}\textbf{tæ} \qquad \text{"on the newspaper"} \tag{41}$$
$$\text{newspaper-LOC}$$

$$\text{qɨrʁɨz-}\textbf{da} \qquad \text{"on the Kyrgyz"} \tag{42}$$
$$\text{Kyrgyz-LOC}$$

In sum, this approach allows us to determine suffix backness by looking only at the final vowel in the stem, which is always specified for backness. This removes the need for a dorsal consonant tier, and allows this pattern to be captured easily by the TSL grammar over T_v described in the previous section.

There are two issues with this approach. The first is that there are a small number of stems in the language that still appear to follow a pattern where vowels are considered before consonants, as exemplified below:

$$\text{tæstiq-}\textbf{tæ} \qquad \text{"on the sanction"} \tag{43}$$
$$\text{sanction-LOC}$$

Here we might expect the final /i/ of the stem to take on the backness of the adjacent uvular, but instead we see that this word takes a front suffix. Words like this are rare, but it is unclear how to achieve the proper backness of the final vowel. One solution is to stipulate that vowels are considered before consonants, which leads back to the problem we hoped to avoid. Another solution is to stipulate that the surface forms of /I/ and /E/ are identical to their underlying forms, and the tendency for these forms to agree in backness with nearby harmonizing vowels and dorsals is simply a coincidence.

The second issue is that there have been no studies looking at the phonetic realization of /i/ and /e/ in stems that differ in the backness of their suffixes. Positing two separate phonemes without such data for the sake of a more theoretically amenable analysis is rather ad hoc, although it has been done. Lindblad, for example, proposes such an analysis, where an underlying contrast between /i/ and /ɨ/ is neutralized by a post-lexical fronting process [30]. This is essentially unfalsifiable. In the following section we present original data from a small study investigating whether there is phonetic support for such an analysis.

5.1   An Acoustic Study of the Transparent Vowel /i/ in Uyghur

In the present study we restrict ourselves to stems containing neither harmonizing vowels nor dorsals. This is because coarticulatory processes, where the articulation of a sound is influenced by nearby sounds, are a confounding factor in trying to show that there is a phonological distinction between /i/ and /ɨ/ in Uyghur. Coarticulatory processes are common across languages, and though they often become phonologized, the presence of coarticulation does not necessarily provide evidence for a phonological process (e.g. [24,36]). In other words, we cannot tell using measurements whether the vowels in (42), for example, are phonologically [ɨ] or simply a little backer because of the nearby uvulars.

Vowel-to-vowel coarticulation has been studied extensively (e.g. [3,11,12]). A finding that is relevant for the pattern described here is that languages vary in their patterns of V-to-V coarticulation: coarticulation tends to be greater when there is less risk of confusion between meaningfully distinct phonemes [31,32]. This suggests that, under the assumption that phonological /ɨ/ and /ə/ do not exist, /i/ and /e/ should be the most susceptible of the Uyghur vowels to coarticulation with nearby back sounds, since these are the only two that have no corresponding vowel differing only in backness.

Less has been said about the effects of uvular consonants on vowels, but studies of languages such as Cochabamba Quechua [15] and Ditidaht [46] show that uvulars produce a backing effect on nearby vowels (particularly front vowels).

These findings suggest that the vowels /i/ and /e/ should exhibit backing around back vowels and uvulars by phonetic coarticulatory processes, and hence acoustic evidence of such cannot be taken as proof of a phonological distinction. A domain that is free from this confound is the set of stems that have neither harmonizing vowels nor dorsal consonants. A significant difference in backness between the vowels in stems that take front suffixes and those that take back suffixes could not be attributed to phonetic coarticulation, and would be compelling evidence for a meaningful phonological contrast. Phonetic studies have been performed for a similar process in Hungarian [5,6], and the methodology here is similar to that employed by Blaho and Szeredi.

Methodology. Two native speakers of Uyghur from the Urumchi region, one male and one female, read the words in Table 6. All words are monosyllabic with only the transparent vowel /i/ and no dorsal consonants. Both speakers were presented with the same list of words. Words were removed if they were unfamiliar to the speaker or if they were produced with no vowel.[8] The speakers did not agree on the backness of all words when suffixes were added: bolded words in the tables highlight disagreements.

Table 6. Word lists for speakers 1 (left) and 2 (right). Bolded forms indicate disagreements in stem backness between the speakers.

Speaker 1:

Front		Back	
/bil/	'know'	/tʃiʃ/	'tooth'
/bir/	'one'	/dil/	'heart'
/biz/	'we'	**/mis/**	**'copper'**
/din/	**'religion'**	/pil/	'elephant'
/iʃ/	'work'	/sirt/	'outside'
/dʒin/	**'Djinn'**	/siz/	'draw'
/min/	'ride'	/til/	'tongue'
/sir/	**'brush'**	/tiz/	'knee'
/siz/	'you'		

Speaker 2:

Front		Back	
/bil/	'know'	/tʃiʃ/	'tooth'
/bir/	'one'	/dil/	'heart'
/biz/	'we'	**/din/**	**'religion'**
/min/	'ride'	/it/	'dog'
/mis/	**'copper'**	**/dʒin/**	**'Djinn'**
/siz/	'you'	/lim/	'beam'
		/pil/	'elephant'
		/pir/	'master'
		/sir/	**'brush'**
		/sirt/	'outside'
		/siz/	'draw'
		/til/	'tongue'
		/tiz/	'knee'

The speakers produced the words in the carrier sentence

tursun hazir	_____	*dɛdi*	
Tursun again	_____	say.PAST	(44)
Tursun said	_____	again	

Words were elicited in two forms: with no harmonizing suffix (bare for nouns, and with the third person past tense suffix -*di* for verbs) and with a harmonizing suffix (the locative -*DA* for nouns, and the infinitive -*mAQ* for verbs, where Q alternates between /k/ and /q/ depending on backness). The purpose of eliciting

---

[8] Uyghur has a process of vowel lenition that can occur adjacent to voiceless consonants: e.g. speaker 1 produced /iʃ/ with a vowel while speaker 2 did not, and speaker 2 produced /it/ with a vowel while speaker 1 did not.

words with a suffix was to confirm the predicted coarticulatory effect of a nearby back or front vowel.

The stem vowels were segmented by hand using Praat [7], and F1 and F2 were extracted at vowel midpoints using a script. We ran two linear mixed effects models in R [40] using the *nlme* package [39], with F1 and F2 as the dependent variables respectively. Stem backness (i.e. whether the stem takes a front or back suffix) and the presence of a harmonizing suffix were the independent variables. Word and subject were random effects.

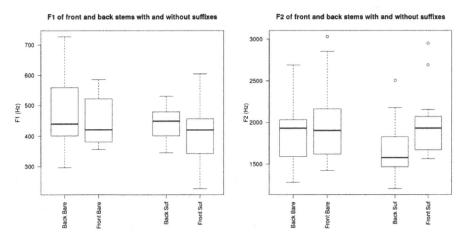

Fig. 2. Plots of F1 (left) and F2 (right)

Results. The results are shown in Fig. 2. There were no significant effects of stem backness or suffix presence on F1. Similarly, there was no significant effect of stem backness on F2, but there was a significant interaction between stem backness and the presence of a suffix on F2 ($\beta = 183.47$; $t = 3.65$; $p < 0.01$). This indicates that back vowels in the suffixes pulled /i/ back, lowering F2. These results provide no evidence that the vowel /i/ provides a cue for backness in stems with no harmonizing vowels or dorsals, though they do demonstrate a coarticulatory effect between /i/ and nearby back vowels.

Discussion. The fact that nearby back vowels induce backing on /i/ suggests that /i/ is indeed susceptible to coarticulation with neighboring segments. Hence it is plausible that in stems containing harmonizing vowels or dorsals, information about these segments may be conveyed through the transparent vowels. The challenge for adopting this analysis is making a convincing case that this is a phonological pattern, not merely a phonetic one. The general results here do not support this analysis: stems containing only the vowel /i/ and non-dorsal consonants show no acoustic distinction based on the backness of their suffixes, and must still be arbitrarily specified as front or back. Although this analysis is able to circumvent the issues for a TSL representation, it does so with no gain in explanatory power, and at the cost of additional stipulation.

6   Uyghur Backness Harmony as a Lexicalized Process

The characterization of Uyghur backness harmony in Sects. 3 and 4 is incompatible with the theory that all phonological stringsets are TSL languages. There is, however, evidence that backness harmony in Uyghur may be learned as a lexical or morphological process, rather than a phonological one, which is consistent with the idea that TSL provides an upper bound on phonological complexity.

Under a morphological account of backness harmony, all stems are specified lexically as taking either front or back suffixes (or, put in a slightly different way, speakers simply memorize which stems take which suffixes). This pattern is easily captured using a TSL grammar, and is consistent with the hypothesis that morphotactic processes are also maximally TSL [1].

Under this analysis, stems lacking harmonizing vowels or dorsals are treated identically to all other stems. What this approach sacrifices is the strong generalization that stems with harmonizing vowels and dorsals tend to take suffixes that agree in backness. Although such approaches run counter to the tendency in generative phonology to limit lexical specification, there is some motivation for adopting them in certain cases. We will present a brief summary of a particularly notorious case, and show that Uyghur satisfies the same motivating criteria.

Positing a morphological process over a phonological one is often based on the complexity of the phonological analysis required to capture the pattern, particularly regarding learnability. Such analyses often require underlying forms that differ substantially from any surface form, baroque interactions between independent processes (e.g. rules or constraints), and some way to capture inconsistent generalizations and variation within or between speakers. Examples include French liaison (e.g. [9]), Polish /o/-/u/ alternation [43], irregular English past tense morphology [2], and possessive prefixes in Odawa [8]. Such cases have two common themes. First, these processes typically originated as predictable and productive phonological patterns that were subsequently obscured by diachronic change. This led to a reanalysis by language learners, since insufficient evidence was available to reconstruct the original pattern. Second, in the absence of reliable structural cues, speakers tend to rely on statistical generalizations to determine the appropriate surface realization in unfamiliar cases.

We focus on Maori passives as a representative example. This was first raised as a challenge for phonological analysis by Hale [19] and has been written on extensively since (see [37] for an excellent overview). Table 7 (from [37]) shows a sample of Maori passive forms.

This pattern developed as the result of all word-final consonants being lost in unsuffixed forms, but maintained in medial position when the passive suffix /-ia/ is present. A phonological analysis must either make reference to properties of the stems that systematically predict particular suffix forms (which are not obvious)

Table 7. Maori passives

Active	Passive	Gloss
/ɸera/	/ɸerahia/	'to spread'
/oma/	/omakia/	'to run'
/inu/	/inumia/	'to drink'
/eke/	/ekeŋia/	'to climb'
/tupu/	/tupuɾia/	'to grow'
/aɸi/	/aɸitia/	'to embrace'
/huna/	/hunaia/	'to conceal'
/kata/	/kataina/	'to laugh'
/ako/	/akona/	'to teach'
/heke/	/hekea/	'to descend'

or assume the presence of word-final consonants underlyingly and a process of surface deletion. Even in the latter case, bizarre assumptions are often required to support this analysis, such as Hale's suggestion that there is an underlying /p/ at the end of certain forms that is never realized in *any* surface form [19]. There is also evidence that these passive forms have been analyzed as separate, competing morphemes by speakers, with /-tia/ coming to be preferred as the default but substantial free variation possible within and between speakers. Sophisticated statistical analyses also show that the suffix form can be predicted reasonably well from subtle properties of the stem, suggesting that speakers may be sensitive to statistical generalizations when choosing the appropriate suffix [37].

Uyghur backness harmony shares many of the properties of the Maori passive system. Disharmonic stems tend to occur mostly in loanwords and compounds [47], suggesting an increase in such stems as more foreign words entered the language. Similarly, as described earlier, historical Uyghur once had a distinction between the front and back vowels /i/-/ɯ/ and /e/-/ɤ/ that collapsed into /i/ and /e/, eliminating the backness contrast that would have determined the suffix of many of the problematic forms discussed here [30]. As demonstrated by the responses of the participants in the study in the previous section, there is also inter-speaker variation on which suffixes certain forms take. We are conducting corpus and experimental studies to evaluate the extent to which the process of backness harmony has been productively acquired by Uyghur speakers [33].

More evidence is needed to establish that Uyghur backness harmony is treated by speakers as a lexicalized process, but there are several points in favor of this account: it provides a consistent analysis, with the generalizations around harmonizing vowels and dorsal consonants reflecting statistical remnants of a past, more reliable stage of the language; it exhibits many of the properties of languages where similar processes have been claimed; and, it is consistent with the theory that TSL provides an approximately correct upper bound on phonological learnability [34]. Assuming that only phonological patterns that can be effectively learned will be effectively propagated, it is perhaps no coincidence that non-TSL patterns are so uncommon in the world's languages.

7 Discussion and Conclusion

A motivation that is commonly put forward for studying phonological processes through the lens of formal complexity is that it allows a detailed characterization of the data that is agnostic to any particular theory. This in turn provides clear requirements for the expressive and restrictive capabilities of any theory leveled at the data (e.g. [21,42]), and also suggests upper bounds on the complexity of patterns that can be effectively learned (e.g. [34]). In this paper we presented a pattern that is a challenge for current hypotheses about how complex phonological stringsets may be.

We first showed that the pattern of stem-suffix backness harmony in Uyghur cannot be generated by a tier-based strictly local grammar, nor by *any* of the subregular language classes previously applied to phonology. This is problematic for the weak subregular hypothesis, which claims that all phonological stringsets are maximally TSL. We then explored an alternative analysis that suggests the pattern is phonological and contained solely on the vowel tier, but rejected it for lack of empirical support. Finally, we showed that even though Uyghur surface strings are beyond TSL, there is some evidence that the pattern of backness harmony may be lexicalized. This is consistent with the idea of TSL as an upper bound on effective phonological learnability. More empirical data is needed to claim with certainty that this is the case, but we feel that this is a promising hypothesis.

We do not go into considerations of the merits of one phonological theory over another, but it is worth noting that the characterization of Uyghur backness harmony bears on this as well. The interaction of backness harmony with other phonological process in Uyghur has been put forward as evidence for a derivational theory of phonology [47], and the considerations discussed here bear on the validity of these claims.

Assuming that backness harmony in Uyghur is governed by the phonotactic knowledge of its speakers, the analysis presented in Sects. 3 and 4 is incompatible with the weak subregular hypothesis. If, on the other hand, backness harmony is indeed a lexicalized process, this would be consistent with the idea that TSL languages provide an approximately correct upper bound on phonological learnability. In addition to the simple presentation of this data as a challenging case for subregular phonology, we hope that we have illustrated how theories of formal complexity can serve as useful conceptual tools in addition to those traditionally employed by phonologists.

Acknowledgements. We would like to thank Tim Hunter, Kie Zuraw, Thomas Graf, two anonymous reviewers, and the attendees of the UCLA phonology seminar for their invaluable feedback. We would also like to thank our Uyghur consultants for sharing their language and culture. Without their generosity and time, none of this would have been possible.

References

1. Aksënova, A., Graf, T., Moradi, S.: Morphotactics as tier-based strictly local dependencies. In: Proceedings of the 14th Annual SIGMORPHON Workshop on Computational Research in Phonetics, Phonology, and Morphology, pp. 121–130 (2016)
2. Albright, A., Hayes, B.: Rules vs. analogy in English past tenses: a computational/experimental study. Cognition **90**, 119–161 (2003)
3. Alfonso, P.J., Baer, T.: Dynamics of vowel articulation. Lang. Speech **25**, 151–173 (1982)
4. Baek, H.: Computational representation of unbounded stress: tiers with structural features. In: Proceedings of the 53rd Annual Meeting of the Chicago Linguistic Society (2017, to appear)
5. Benus, S., Gafos, A.: Articulatory characteristics of Hungarian 'transparent' vowels. J. Phon. **35**, 271–300 (2007)
6. Blaho, S., Szeredi, D.: Hungarian neutral vowels: a microcomparison. Nordlyd **40**, 20–40 (2013)
7. Boersma, P., Weenink, D.: Praat: doing phonetics by computer [Computer Program]. Version 6.0.37. http://www.praat.org/. Accessed 17 Feb 2018
8. Bowers, D.: A system for morphophonological learning and its implications for language change. Ph.D. thesis, UCLA (2015)
9. Bybee, J.: Frequency effects on French liaison. In: Bybee, J., Hopper, P. (eds.) Frequency and the Emergence of Linguistic Structure, pp. 337–359. John Benjamins, Amsterdam (2001)
10. Chandlee, J.: Strictly local phonological processes. Ph.D. thesis, The University of Delaware (2014)
11. Choi, J.-D., Keating, P.A.: Vowel-to-vowel coarticulation in Slavic languages. J. Acoust. Soc. Am. **88**, S54 (1990)
12. Cole, J., Linebaugh, G., Munson, C., McMurray, B.: Unmasking the acoustic effects of vowel-to-vowel coarticulation: a statistical modeling approach. J. Phon. **38**, 167–184 (2010)
13. de Santo, A., Graf, T.: Structure sensitive tier projection: applications and formal properties. Ms., Stony Brook University (2017)
14. Eilenberg, S.: Automata, Languages, and Machines. Academic Press Inc., Cambridge (1974)
15. Gallagher, G.: Vowel height allophony and dorsal place contrasts in Cochabamba Quechua. Phonetica **73**, 101–119 (2016)
16. Gold, E.M.: Language identification in the limit. Inf. Control **10**, 447–474 (1967)
17. Graf, T.: Comparing incomparable frameworks: a model theoretic approach to phonology. In: University of Pennsylvania Working Papers in Linguistics, vol. 16 (2010)
18. Graf, T.: The power of locality domains in phonology. Phonology **34**, 385–405 (2017)
19. Hale, K.: Review of Hohepa (1967). J. Polyn. Soc. **77**, 83–99 (1968)
20. Heinz, J.: Learning long-distance phonotactics. Linguist. Inq. **41**, 623–661 (2010)
21. Heinz, J.: The computational nature of phonological generalizations. In: Phonological Typology (2018)
22. Heinz, J., Kasprzik, A., Kötzing, T.: Learning with lattice-structure hypothesis spaces. Theoret. Comput. Sci. **457**, 111–127 (2012)
23. Heinz, J., Rawal, C., Tanner, H.G.: Tier-based strictly local constraints for phonology. In: Proceedings of the 49th Annual Meeting of the Association for Computational Linguistics: Shortpapers, pp. 58–64 (2011)

24. Hyman, L.M.: Phonologization. In: Julliand, A. (ed.) Linguistic Studies Offered to Joseph Greenberg. volume 2, pp. 407–418. Anma Libri, Saratoga (1976)
25. Jardine, A.: Computationally, tone is different. Phonology **33**, 247–283 (2016)
26. Jardine, A., Heinz, J.: Learning tier-based strictly 2-local languages. Trans. ACL **4**, 87–98 (2016)
27. Jardine, A., McMullin, K.: Efficient learning of tier-based strictly k-local languages. In: Drewes, F., Martín-Vide, C., Truthe, B. (eds.) LATA 2017. LNCS, vol. 10168, pp. 64–76. Springer, Cham (2017). https://doi.org/10.1007/978-3-319-53733-7_4
28. Johnson, C.D.: Formal Aspects of Phonological Decription. Mouton, The Hague (1972)
29. Kaplan, R., Kay, M.: Regular models of phonological rule systems. Comput. Linguist. **20**, 331–378 (1994)
30. Lindblad, V.M.: Neutralization in Uyghur. University of Washington, Seattle (1990)
31. Manuel, S.Y.: The role of contrast in limiting vowel-to-vowel coarticulation in different languages. J. Acoust. Soc. Am. **88**, 1286–1298 (1990)
32. Manuel, S.Y., Krakow, R.A.: Universal and language particular aspects of vowel-to-vowel coarticulation. Haskins Lab. Status Rep. Speech Res. **77**(78), 69–78 (1984)
33. Mayer, C., Major, T.: On the productivity of Uyghur backness harmony (in progress)
34. McMullin, K.: Tier-based locality in long-distance phonotactics: learnability and typology. Ph.D. thesis, University of British Columbia (2016)
35. McNaughton, R., Papert, S.: Counter-Free Automata. MIT Press, Cambridge (1971)
36. Ohala, J.: The listener as a source of sound change. In: Masek, C.S., Hendrick, R.A., Miller, M.F. (eds.) Papers from the Parasession on Language and Behavior, pp. 178–203. Chicago Linguistic Society, Chicago (1981)
37. Jones, 'Ō.P.: Phonotactic probability and the Māori passive. In: Proceedings of the Tenth Meeting of the ACL Special Interest Group on Computational Morphology and Phonology, pp. 39–48 (2008)
38. Pin, J.E.: Varieties of Formal Languages. Plenum Publishing Co., New York (1986)
39. Pinheiro, J., Bates, D., DebRoy, S., Sarkar, D., R Core Team: nlme: Linear and nonlinear mixed effects models. R Package Version 3.1-131 (2017)
40. R Core Team: R: A Language and Environment for Statistical Computing. R Foundation for Statistical Computing, Vienna, Austria (2017)
41. Rogers, J., Heinz, J., Fero, M., Hurst, J., Lambert, D., Wibel, S.: Cognitive and subregular complexity. In: Morrill, G., Nederhof, M.-J. (eds.) FG 2012–2013. LNCS, vol. 8036, pp. 90–108. Springer, Heidelberg (2013). https://doi.org/10.1007/978-3-642-39998-5_6
42. Rogers, J., Pullum, G.K.: Aural pattern recognition experiments and the subregular hierarchy. J. Log. Lang. Inf. **20**, 329–342 (2011)
43. Sanders, R.N.: Opacity and sound change in the Polish lexicon. Ph.D. thesis, UCSC (2003)
44. Schützenberger, M.P.: On finite monoids having only trivial subgroups. Inf. Control **8**, 190–194 (1965)
45. Simon, I.: Piecewise testable events. In: Brakhage, H. (ed.) GI-Fachtagung 1975. LNCS, vol. 33, pp. 214–222. Springer, Heidelberg (1975). https://doi.org/10.1007/3-540-07407-4_23
46. Sylak-Glassman, J.: The effects of post-velar consonants on vowels in Ditidaht. In: Weber, N., Sadlier-Brown, E., Guntly, E. (eds.) Papers for the International

Conference on Salish and Neighbouring Languages 49, vol. 37. University of British Columbia Working Papers in Linguistics (2014)
47. Vaux, B.: Disharmony and derived transparency in Uyghur vowel harmony. In: Proceedings of NELS, vol. 30, pp. 671–698 (2000)
48. Yli-Jyrä, A.: Describing syntax with star-free regular expressions. In: EACL (2003)
49. Yli-Jyrä, A.M., et al.: On finite-state tonology with autosegmental representations. In: Proceedings of the 11th International Conference on Finite State Methods and Natural Language Processing. Association for Computational Linguistics (2013)

Bracket Induction for Lambek Calculus with Bracket Modalities

Glyn Morrill[1], Stepan Kuznetsov[2,5]($\boxtimes$), Max Kanovich[3,5], and Andre Scedrov[4,5]

[1] Universitat Politècnica de Catalunya, Barcelona, Spain
morrill@cs.upc.edu
[2] Steklov Mathematical Institute of the RAS, Moscow, Russia
sk@mi.ras.ru
[3] University College London, London, UK
m.kanovich@ucl.ac.uk
[4] University of Pennsylvania, Philadelphia, USA
scedrov@math.upenn.edu
[5] National Research University Higher School of Economics, Moscow, Russia

Abstract. Relativisation involves dependencies which, although unbounded, are constrained with respect to certain island domains. The Lambek calculus **L** can provide a very rudimentary account of relativisation limited to unbounded peripheral extraction; the Lambek calculus with bracket modalities **Lb** can further condition this account according to island domains. However in naïve parsing/theorem-proving by backward chaining sequent proof search for **Lb** the bracketed island domains, which can be indefinitely nested, have to be specified in the linguistic input. In realistic parsing word order is given but such hierarchical bracketing structure cannot be assumed to be given. In this paper we show how parsing can be realised which induces the bracketing structure in backward chaining sequent proof search with **Lb**.

Keywords: Lambek calculus with brackets · Bracket induction
Categorial grammar

1 Introduction

Relativisation involves dependencies which can be medial as well as peripheral and which, although unbounded, are constrained with respect to certain island domains; furthermore these unbounded dependencies can be multiple, or parasitic, in a way which appears to depend on islands. The Lambek calculus **L** of Lambek [9] can provide a very rudimentary account of relativisation limited to unbounded peripheral extraction; the Lambek calculus with bracket modalities **Lb** of Morrill [13] and Moortgat [11] can further condition this account according to island domains; and the Lambek calculus with bracket modalities and universal subexponential **Lb!** (after Girard [3]) accommodates furthermore medial and parasitic extraction (Morrill [14]). However in naïve parsing/theorem-proving by backward chaining sequent proof search for **Lb** and **Lb**! the bracketed island domains, which can be indefinitely nested, have to be specified in the linguistic input. In realistic parsing word order is given but such hierarchical bracketing structure cannot be assumed to be given. In this paper we show how parsing can be realised which induces the bracketing structure in backward chaining sequent proof search for **Lb**.

1.1   Relativisation

Relativisation is an unbounded dependency construction; the distance between a relative pronoun (filler) and the position it binds (gap) can be unboundedly long:

(1) a. the man that$_i$ Mary loves e_i
      b. the man that$_i$ John thinks Mary loves e_i
      c. the man that$_i$ Suzy knows John thinks Mary loves e_i
      $\vdots$

Some domains are islands to relativisation and cannot be penetrated by the filler-gap (extraction) dependency, for example adverbial phrases are weak islands (extraction is semi-acceptable) and relative clauses themselves are strong islands (extraction is unacceptable):

(2) a. ?the paper that$_i$ John laughed [without reading e_i]
      b. (?)the paper that$_i$ John went to Paris [without reading e_i]
      c. *the waitress that$_i$ John saw the man [that married e_i]

Relativisation can be medial:

(3) the contract that$_i$ John signed e_i yesterday

And although islands block singleton extractions, relativisation can have a parasitic gap in a weak island dependent on a non-island host gap:

(4) the paper that$_i$ John filed e_i [without reading e_i]

Such parasitic gaps can also appear in subjects, which are weak islands:

(5) a. ??the man that$_i$ [the friends of e_i] laughed
      b. the man that$_i$ [the friends of e_i] praised e_i

A single host gap can license parasitic gaps in multiple islands; for example:

(6) the paper that$_i$ [the editor of e_i] filed e_i [without reading e_i]

In this paper we give an account in terms of the Lambek calculus with bracket modalities **Lb** of the data of (1–2). We provide a calculus and consider a bracket-inducing parsing/theorem-proving algorithm. We illustrate this algorithm on an example of a lexical grammar for a small fragment of English. The input of this algorithm is just a string (linearly ordered sequence of words), without any bracketing information. The job of the algorithm is to induce (guess) the correct placement of brackets, as well as to derive the resulting **Lb** sequent. The algorithm relies on the assignment of types drawn from the lexical grammar. In Sect. 6 we discuss the complexity of our algorithm, in comparison with both a naïve approach with brute-force guessing of the correct bracketing and the pseudo-polynomial algorithm for theorem-proving (but not parsing) in **Lb** presented in our earlier paper [5].

As for examples (3–6), their parsing involves **Lb!**, the extension of **Lb** with the universal exponential modality, which is beyond the scope of the present paper. For **Lb!**, the derivability problem in general is algorithmically undecidable [7]. A practically useful fragment, however, guarded by the so-called *bracket non-negative condition,* was shown to be decidable [17] and to belong to the NP complexity class [7]. In other words, despite undecidability of the whole calculus, practical parsing with **Lb!** has the same complexity as for **Lb** without the exponential. We hope to extend the results presented here from **Lb** to **Lb!** (restricted by the bracket non-negative condition) in a subsequent paper.

Our analysis of linguistic examples in this paper follows Morrill [14,20]. This paper's purposes are mostly technical. Namely, we present an approach that allows the parsing algorithm for **Lb** to induce open and closed "symbolic" brackets by itself and therefore avoid requesting this information (which is not part of the actual text in natural language) from the user. Therefore, we refrain from deep discussions of the design of the lexical grammar itself and its empirical justification. In particular, we leave beyond the scope of this paper the subtle issues of *semi-grammaticality* [20, Sect. 5.4] of extraction from weak islands; see, for instance, examples (2b) and (5a). For such examples, Morrill [20, Sect. 5.4] suggests including further structural rules that allow some violation of bracketing, but with a cost for such. The number of applications of such rules is supposed to be counted, and the more times they are used the less grammatical the target sentence is considered. The modification of the algorithm described in this paper to allow such rules is a topic for further investigation.

1.2 Lambek Calculus with Bracket Modalities

The set **Tp** of types of the Lambek calculus with bracket modalities **Lb** is defined in terms of a set $\mathcal{P}$ of primitive types as follows:

$$\mathbf{Tp} ::= \mathcal{P} \mid \langle\rangle\mathbf{Tp} \mid [\,]^{-1}\mathbf{Tp} \mid \mathbf{Tp}{\bullet}\mathbf{Tp} \mid \mathbf{Tp}\backslash\mathbf{Tp} \mid \mathbf{Tp}/\mathbf{Tp}$$

A configuration **Config** is a well-bracketed string of types:

$$\textbf{Config} \quad ::= \textbf{TreeTerm} \mid \textbf{Config}, \textbf{Config}$$
$$\textbf{TreeTerm} ::= \textbf{Tp} \mid [\textbf{Config}]$$

Note that this definition builds in "Lambek's restriction" whereby configurations (and bracketed configurations) are non-empty. Lambek's restriction, even in the fragment without brackets, is motivated linguistically; otherwise we could derive the sequent $(CN/CN)/(CN/CN), CN \Rightarrow CN$, validating grammatically incorrect phrases like "very book" (compare with "very interesting book", parsed as $(CN/CN)/(CN/CN), CN/CN, CN \Rightarrow CN$). In this paper, Lambek's restriction is crucial for the construction to work (see Sect. 5).

A sequent is an expression of the form **Config** $\Rightarrow$ **Tp**.

1.3 Sequent Calculus for Lb

Let $\Delta(\Gamma)$ signify a structure Δ with a distinguished substructure Γ. The sequent calculus for **Lb** is as follows:

$$\frac{}{P \Rightarrow P}\, id \qquad \frac{\Gamma \Rightarrow A \qquad \Delta(A) \Rightarrow B}{\Delta(\Gamma) \Rightarrow B}\, Cut$$

$$\frac{\Delta(A) \Rightarrow B}{\Delta([[]^{-1}A]) \Rightarrow B}\, []^{-1}L \qquad \frac{[\Delta] \Rightarrow B}{\Delta \Rightarrow []^{-1}B}\, []^{-1}R$$

$$\frac{\Delta([A]) \Rightarrow B}{\Delta(\langle\rangle A) \Rightarrow B}\, \langle\rangle L \qquad \frac{\Delta \Rightarrow B}{[\Delta] \Rightarrow \langle\rangle B}\, \langle\rangle R$$

$$\frac{\Delta(A, B) \Rightarrow D}{\Delta(A \bullet B) \Rightarrow D}\, \bullet L \qquad \frac{\Delta \Rightarrow A \qquad \Gamma \Rightarrow B}{\Delta, \Gamma \Rightarrow A \bullet B}\, \bullet R$$

$$\frac{\Gamma \Rightarrow B \qquad \Delta(C) \Rightarrow D}{\Delta(C/B, \Gamma) \Rightarrow D}\, /L \qquad \frac{\Gamma, B \Rightarrow C}{\Gamma \Rightarrow C/B}\, /R$$

$$\frac{\Gamma \Rightarrow A \qquad \Delta(C) \Rightarrow B}{\Delta(\Gamma, A\backslash C) \Rightarrow B}\, \backslash L \qquad \frac{A, \Gamma \Rightarrow C}{\Gamma \Rightarrow A\backslash C}\, \backslash R$$

Moortgat [11] shows that this calculus enjoys Cut-elimination: that every theorem has a Cut-free proof. We omit Cut in what follows.

ate: $(\langle\rangle N\backslash S)/N$
built: $(\langle\rangle N\backslash S)/N$
cat: CN
contract: CN
dog: CN
editor: CN/PP
filed: $(\langle\rangle N\backslash S)/N$
friends: CN/PP
house: CN
in: $[]^{-1}(((\langle\rangle N\backslash S)\backslash(\langle\rangle N\backslash S))/N$
Jack: N
John: N
killed: $(\langle\rangle N\backslash S)/N$
knows: $(\langle\rangle N\backslash S)/S$
laughed: $\langle\rangle N\backslash S$
lay: $\langle\rangle N\backslash S$
loves: $(\langle\rangle N\backslash S)/N$
malt: CN
man: CN
married: $(\langle\rangle N\backslash S)/N$
Mary: N
paper: CN
Paris: N
praised: $(\langle\rangle N\backslash S)/N$
rat: CN
reading: $(\langle\rangle N\backslash S)/N$
saw: $(\langle\rangle N\backslash S)/N$
signed: $(\langle\rangle N\backslash S)/N$
Suzy: N
that: $[]^{-1}[]^{-1}(CN\backslash CN)/(\langle\rangle N\backslash S)$
that: $[]^{-1}[]^{-1}(CN\backslash CN)/(S/N)$
the: N/CN
thinks: $(\langle\rangle N\backslash S)/S$
to: PP/N
waitress: CN
went: $(\langle\rangle N\backslash S)/PP$
without: $[]^{-1}((\langle\rangle N\backslash S)\backslash(\langle\rangle N\backslash S))/(\langle\rangle N\backslash S)$
worried: $(\langle\rangle N\backslash S)/N$
yesterday: $(\langle\rangle N\backslash S)\backslash(\langle\rangle N\backslash S)$

Fig. 1. Lexical grammar

1.4 Grammar

Consider the micro-**Lb** lexical grammar in Fig. 1 whereby weak islands are singly
bracketed and strong islands doubly bracketed. For example, (1a) is derived as
follows:

$$\cfrac{\cfrac{\cfrac{N \Rightarrow N}{[N] \Rightarrow \langle\rangle N}\langle\rangle R \quad S \Rightarrow S}{\cfrac{[N], \langle\rangle N\backslash S \Rightarrow S}{}}\backslash L}{\cfrac{N \Rightarrow N \qquad [N], (\langle\rangle N\backslash S)/N, N \Rightarrow S}{[N], (\langle\rangle N\backslash S)/N \Rightarrow S/N}/R}/L$$

$$\cfrac{\cfrac{\cfrac{CN \Rightarrow CN \quad CN \Rightarrow CN}{CN, CN\backslash CN \Rightarrow CN}\backslash L}{CN, [[]^{-1}(CN\backslash CN)] \Rightarrow CN}[]^{-1}L}{CN, [[[]^{-1}[]^{-1}(CN\backslash CN)]] \Rightarrow CN}[]^{-1}L$$

$$\cfrac{}{CN, [[[]^{-1}[]^{-1}(CN\backslash CN)/(S/N), [N], (\langle\rangle N\backslash S)/N]] \Rightarrow CN}/L$$

Examples (2a–c) are blocked because the bracket modalities require the island bracketings which prevent the hypothetical subtype of the relative pronoun from associating into the object positions of "reading" and "married" respectively. We discuss example (2a) in detail in Subsect. 5.3.

2   Inducing Brackets

The general strategy will be to represent symbolic half-brackets in the input by variables for which the flow of information is propagation from endsequent to axiom leaves and instantiation from axiom leaves to endsequent. Thus we need a coding of brackets which represents the terminal yield of sequents in terms of open and closed "symbolic" half-brackets which are uninstantiated bottom up.

Such antecedents are of the pattern:

$$(\text{Onset } \mathbf{Tp} \text{ Offset})^+$$

where Onset is of the form [*, **Tp** is a type, and Offset is of the form]*. This formulation builds in Lambek's restriction whereby there must be at least one type in an antecedent and within brackets.

We represent Onsets and Offsets by lists of zeros; the length of the list is the number of brackets. For example, [[[is coded $[0, 0, 0]$,] is coded $[0]$ and no brackets are coded by the empty list $[]$ $(\emptyset)$.[1]

A symbolically bracketed antecedent Δ is *well-bracketed* (is a BI-**Config**), if it is the case that at every point of Δ, the sum of all Onsets to the left of this point is greater or equal than the sum of all Offsets to the left of this point, and that for the whole Δ these two sums are equal.

In the following section we present the bracket inducing rules.

3   Bracket Inducing Rules

Regarding notation, in bracket inducing BI-**Lb** rules, we use Δ (and also Δ_1, Δ_2) for sequences starting with an Onset F and ending with an Offset G, and

---

[1] We use this notation because it prefigures a planned future extension to exponentials in which onsets are coded by lists of naturals representing the size of so-called stoups; the zeros in the coding of **Lb** essentially mean that here all stoups are empty.

Ω for sequences starting and ending with types (in particular, just a singleton type is also an Ω).

All rules have a premise-to-conclusion property by which all sequents generated have well-bracketed antecedents. Fragments of the antecedent, however, are not necessarily well-bracketed: in a sequent of the form $\Delta_1, \Delta, \Delta_2 \Rightarrow B$ an opening bracket in Δ_1 could have its corresponding closing bracket in Δ_2, and in this case Δ_1 and Δ_2 are not well-bracketed. Also recall that we are working with a Cut-free version of **Lb**, therefore the Cut rule is not included in the BI-**Lb** calculus.

Identity axiom:

$$\frac{}{\emptyset, P, \emptyset \Rightarrow P} \, id$$

Bracket modalities:

$$\frac{\Delta_1, F, A, G, \Delta_2 \Rightarrow B}{\Delta_1, F \oplus [0], []^{-1}A, [0] \oplus G, \Delta_2 \Rightarrow B} \, []^{-1}L \qquad \frac{[0] \oplus F, \Omega, G \oplus [0] \Rightarrow B}{F, \Omega, G \Rightarrow []^{-1}B} \, []^{-1}R$$

$$\frac{\Delta_1, F \oplus [0], A, [0] \oplus G, \Delta_2 \Rightarrow B}{\Delta_1, F, \langle \rangle A, G, \Delta_2 \Rightarrow B} \, \langle \rangle L \qquad \frac{F, \Omega, G \Rightarrow B}{[0] \oplus F, \Omega, G \oplus [0] \Rightarrow \langle \rangle B} \, \langle \rangle R$$

Lambek connectives: the original Lambek rules should also be modified in order to include the new bracket induction mechanism, as shown below.

$$\frac{\Delta_1, F, A, \emptyset, \emptyset, B, G, \Delta_2 \Rightarrow D}{\Delta_1, F, A \bullet B, G, \Delta_2 \Rightarrow D} \, \bullet L \qquad \frac{\Delta_1 \Rightarrow A \quad \Delta_2 \Rightarrow B}{\Delta_1, \Delta_2 \Rightarrow A \bullet B} \, \bullet R$$

$$\frac{F_1, \Omega, G_1 \Rightarrow B \quad \Delta_1, F_2, C, G_2, \Delta_2 \Rightarrow D}{\Delta_1, F_2, C/B, \emptyset, F_1, \Omega, G_1 \oplus G_2, \Delta_2 \Rightarrow D} \, /L \qquad \frac{\Delta, \emptyset, B, \emptyset \Rightarrow C}{\Delta \Rightarrow C/B} \, /R$$

$$\frac{F_1, \Omega, G_1 \Rightarrow A \quad \Delta_1, F_2, C, G_2, \Delta_2 \Rightarrow D}{\Delta_1, F_2 \oplus F_1, \Omega, G_1, \emptyset, A \backslash C, G_2, \Delta_2 \Rightarrow D} \, \backslash L \qquad \frac{\emptyset, A, \emptyset, \Delta \Rightarrow C}{\Delta \Rightarrow A \backslash C} \, \backslash R$$

4   Correctness

We define a translation $\# : \mathbf{Config} \to$ BI-$\mathbf{Config}$ in terms of injective functions $\#(0^*, 0^*) : \mathbf{Config} \to$ BI-$\mathbf{Config}$ as follows:

(7) a. $\#(\Delta) = \#(\emptyset, \emptyset)(\Delta)$
     b. $\#(F, G)(P) = F, P, G$
        $\#(F, G)([\Delta]) = \#(F \oplus [0], [0] \oplus G)(\Delta)$
        $\#(F, G)(\Gamma, \Delta) = \#(F, \emptyset)(\Gamma), \#(\emptyset, G)(\Delta)$

Then we have the following:

(8) **Proposition** (*BI completeness*)

$$\vdash_{\mathbf{Lb}} \Gamma \Rightarrow A \quad \Longrightarrow \quad \vdash_{\text{BI-}\mathbf{Lb}} \#(\Gamma) \Rightarrow A$$

Proof. Trivial induction on derivation in **Lb**.   Q.E.D.

(9) **Lemma** (*BI soundness*)

$$\vdash_{\mathbf{Lb}} \Gamma \Rightarrow A \quad \Longleftarrow \quad \vdash_{\text{BI-}\mathbf{Lb}} \#(\Gamma) \Rightarrow A$$

Proof. First we show, by induction on derivation in BI-**Lb**, that if $\vdash_{\text{BI-}\mathbf{Lb}}$ $\Delta \Rightarrow A$, then $\Delta = \#(\Gamma)$ for some $\Gamma \in$ **Config**; in particular, that Δ is well-bracketed in the sense of Sect. 2. In the only non-trivial induction steps, those of $/L$ and $\backslash L$, the well-bracketedness of the antecedent in the conclusion follows from the fact that the antecedent of the minor (left) premise is contiguous in the antecedent of the conclusion. Second, we prove the soundness lemma itself, also by induction on the derivation of $\#(\Gamma) \Rightarrow A$ in BI-**Lb**. Again, the non-trivial case here is the branching rule, $/L$, for example. Here we use the above consideration to establish the fact that both premises are actually #-translations of some **Lb** sequents. After that, the instance of $/L$ in BI-**Lb** transforms into an instance of the corresponding rule in **Lb**.   Q.E.D.

(10) **Theorem** (*BI correctness*)

$$\vdash_{\mathbf{Lb}} \Gamma \Rightarrow A \quad \Longleftrightarrow \quad \vdash_{\text{BI-}\mathbf{Lb}} \#(\Gamma) \Rightarrow A$$

Proof. By BI soundness and BI completeness.   Q.E.D.

Notice that throughout this section F_i's and G_i's in BI-**Lb** derivations are *ground terms* of the form $[0, 0, \ldots, 0]$ or $\emptyset$ (representing *constant* natural numbers). Thus, BI-**Lb** is actually just an equivalent formulation of **Lb**, and no real bracket induction is taking place yet. In the next section, we treat F_i's and G_i's as *variables,* whose values are not yet known when we start the proof search.

5   Parsing

5.1   Bracket-Inducing Proof-Search and Parsing

The usual parsing procedure using a categorial grammar works as follows. Using a lexical grammar, such as that of Fig. 1, we assign types to words of a string, and these types form the left-hand side of the sequent we are going to derive. The right-hand side is a fixed type, usually primitive, like S for "sentence," for example. If the sequent is derivable, the string is considered valid, and, moreover, we can extract some semantic information from the proof via Curry–Howard correspondence (see [12]). This works perfectly well when antecedents are just linearly ordered lists of formulae (for example, with the "pure" Lambek calculus). With brackets, the situation becomes more involved. Now left-hand sides of sequents also include the bracketing structure. In the naïve generalisation of the Lambek-style parsing procedure to **Lb**-grammars, the bracketing structure should be provided to the parsing algorithm along with the input string and the lexical grammar. For example, since in our lexicon transitive verbs have type $(\langle\rangle N\backslash S)/N$ (the subject forms a weak island), one needs to provide "[John] loves Mary" instead of just "John loves Mary" as the input string for the algorithm. The parser/theorem-prover CatLog3 of Morrill [15] currently depends on being provided with such bracketing structure in the input. A real natural language string, however, does not come with any brackets. Thus, the more appropriate formulation of the parsing problem involves an existential quantifier over possible bracketings. A string is considered valid if the sequent constructed from the corresponding types is derivable in **Lb** for *some* bracketing.

More formally, a string s is of type A according to a grammar if and only if it has a factorization $s = w_1 + w_2 + \cdots + w_n$ such that $w_1: A_1, w_2: A_2, \ldots$, and $w_n: A_n$ are in the grammar lexicon and the following is derivable in BI-**Lb**:

(11) $F_1, A_1, G_1, F_2, A_2, G_2, \ldots, F_n, A_n, G_n \Rightarrow A$

for *some* values of F_1, G_1, F_2, G_2, ..., F_n, G_n. (Recall that F_i and G_i are natural numbers, but written in the form $[0, 0, \ldots, 0]$, linearly bounded, for a given lexicon, by the total length of the sequent.)

Notice that Lambek's non-emptiness restriction is crucial here. It guarantees that every pair of brackets has at least one formula inside, thus brackets are well-organised: in the beginning of the sequent there is $[[\ldots[$ (corresponding to F_1), in the end $]\ldots]]$ (corresponding to G_n), and between two formulae $]\ldots][\ldots[$ (corresponding to G_i, F_{i+1}). Without Lambek's restriction, a more weird behaviour is possible. For example, the sequent $s/\langle\rangle(p/p) \to s$ becomes derivable, but only with the following bracketing: $s/\langle\rangle(p/p), [] \to s$, which does not map to a BI sequent of the form (11), and thus would not be found by the algorithm we describe below.

The proof search procedure using bracket induction works as follows. We start with a sequent with no brackets placed, $A_1, \ldots, A_n \Rightarrow A$, and insert *variables* for symbolic brackets: $F_1, A_1, G_1, F_2, A_2, G_2, \ldots, F_n, A_n, G_n \Rightarrow A$. Then we do proof search from the goal sequent to axiom leaves, annotating each rule application with *side effects*, which are equations on F_i's and G_i's. For each new sequent, we introduce new fresh variables in the places where bracketing is altered, or put $\emptyset$, where the rule postulates that there should be no bracketing. Symbolic brackets in the context are just copied upwards. The rules of BI-**Lb** are annotated with side effects as follows (side effect annotations are placed on the right of the rules):

Identity axiom:
$$\frac{F = \emptyset}{F, P, G \Rightarrow P} \ G = \emptyset$$

Rules for bracket modalities:

$$\frac{\Delta_1, F', A, G', \Delta_2 \Rightarrow B}{\Delta_1, F, []^{-1}A, G, \Delta_2 \Rightarrow B} \ \begin{array}{l} F = F' \oplus [0] \\ G = G' \oplus [0] \end{array} \qquad \frac{F', \Omega, G' \Rightarrow B}{F, \Omega, G \Rightarrow []^{-1}B} \ \begin{array}{l} F \oplus [0] = F' \\ G \oplus [0] = G' \end{array}$$

$$\frac{\Delta_1, F', A, G', \Delta_2 \Rightarrow B}{\Delta_1, F, \langle\rangle A, G, \Delta_2 \Rightarrow B} \ \begin{array}{l} F \oplus [0] = F' \\ G \oplus [0] = G' \end{array} \qquad \frac{F', \Omega, G' \Rightarrow B}{F, \Omega, G \Rightarrow \langle\rangle B} \ \begin{array}{l} F = F' \oplus [0] \\ G = G' \oplus [0] \end{array}$$

Lambek rules:

$$\frac{\Delta_1, F, A, \emptyset, \emptyset, B, G, \Delta_2 \Rightarrow D}{\Delta_1, F, A\bullet B, G, \Delta_2 \Rightarrow D} \qquad \frac{\Delta_1 \Rightarrow A \qquad \Delta_2 \Rightarrow B}{\Delta_1, \Delta_2 \Rightarrow A\bullet B}$$

$$\frac{F_1, \Omega, G'_1 \Rightarrow B \qquad \Delta_1, F_2, C, G'_2, \Delta_2 \Rightarrow D}{\Delta_1, F_2, C/B, G_1, F_1, \Omega, G_2, \Delta_2 \Rightarrow D} \ \begin{array}{l} G_1 = \emptyset \\ G_2 = G'_1 \oplus G'_2 \end{array} \qquad \frac{\Delta, \emptyset, B, \emptyset \Rightarrow C}{\Delta \Rightarrow C/B}$$

and symmetrically for \. The rules without annotations have no side effects.

The proof search procedure yields a tree which we call *pre-derivation*. In the pre-derivation, instead of constant ground terms (natural numbers) we use variables or $\emptyset$'s. On the other hand, the pre-derivation is annotated by side-effect equations that allow computing the ground terms for symbolic brackets.

The side effect equations are actually very simple: on the left-hand side we have either a term with only one occurrence of a symbolic bracket variable from the conclusion of the rule application, or just a ground term, if there was a $\emptyset$. The right-hand side includes variables from the premises. Thus, once the tree is constructed upto axioms, the algorithm tries to resolve side effects going backwards (from axiom leaves to the goal sequent). For axiom leaves, the symbolic bracket variables receive the zero ($\emptyset$) value, and then we recursively go down. At each step we either evaluate the new variable in the conclusion, or, if there was a $\emptyset$, check whether the right-hand side of the equation (which is already computed) is also $\emptyset$.

Sometimes the side effect equations could be non-satisfiable. For example, for the sequent $P \Rightarrow []^{-1}P$, which is not derivable under any bracket assignment, we have $F_1, P, G_1 \Rightarrow []^{-1}P$, and after applying $[]^{-1}R$ (which is the only possible rule here) we get $F_2, P, G_2 \Rightarrow P$ with side effects $F_2 = F_1 \oplus [0]$ and $G_2 = F_1 \oplus [0]$. On the other side, we have $F_2 = G_2 = \emptyset$ from the axiom, which gives a non-satisfiable equation $F_1 \oplus [0] = \emptyset$ (recall that F_i and G_i should always be non-negative integers).

Another, more sophisticated and linguistically relevant example is given in Subsect. 5.3.

Therefore, even when a pre-derivation is obtained, we still have to resolve the side-effects; fortunately, this can be done in linear time and does not substantially slow down the proof search process. If resolving of side effects succeeds, variables get replaced with ground terms (natural numbers), obtaining a derivation (in the standard sense) in BI-**Lb**. By Theorem (10), this yields derivability of the original sequent, with some brackets assigned, in **Lb**. If resolving side-effects failed, the algorithm continues the proof search.

Correctness of the bracket-inducing proof search algorithm is justified by the following theorem, whose proof is straightforward.

(12) **Theorem.** For a sequent of the form $F_1, A_1, G_1, \ldots, F_n, A_n, G_n \Rightarrow A$, in which F_i and G_j are variables, the algorithm described above yields all BI-**Lb** derivations of all instances of this sequents with ground terms (natural numbers) substituted for these variables. (In particular, it returns "not derivable" if there are no such derivations.)

5.2 A Positive Example

In this subsection we run our bracket-inducing proof search algorithm on the common noun group in (1a), "man that Mary loves." In order to show that it is of type CN, if we prove the following in the BI-**Lb** calculus:

$$(13) \quad \begin{array}{l} F_1, CN, G_1, F_2, []^{-1}[]^{-1}(CN \backslash CN)/(S/N), G_2, F_3, N, G_3, \\ F_4, (\langle \rangle N \backslash S)/N, G_4 \Rightarrow CN \end{array}$$

The following pre-derivation is annotated with side effects at axiom leaves and rules. The goal sequent (13) above is pre-derived using $/L$ from two sequents with the following pre-derivations (we omit side effects that are trivially satisfied, like $\emptyset = \emptyset \oplus \emptyset$):

$$
\cfrac{
  \cfrac{
    \cfrac{
      \cfrac{\rule{2cm}{0.4pt}}{F_{11}, N, G_9 \Rightarrow N} \begin{array}{l} F_{11} = \emptyset \\ G_9 = \emptyset \end{array}
    }{F_5, N, G_3 \Rightarrow \langle \rangle N} \begin{array}{l} F_5 = [0] \oplus F_{11} \\ G_3 = G_9 \oplus [0] \end{array}
    \quad
    \cfrac{\rule{2cm}{0.4pt}}{F_6, S, \emptyset \Rightarrow S} \begin{array}{l} F_6 = \emptyset \\ F_4 = \emptyset \\ F_3 = F_6 \oplus F_5 \end{array}
  }{
    \cfrac{\emptyset, N, \emptyset \Rightarrow N \qquad F_3, N, G_3, F_4, \langle \rangle N \backslash S, \emptyset \Rightarrow S}{F_3, N, G_3, F_4, (\langle \rangle N \backslash S)/N, G_5, \emptyset, N, \emptyset \Rightarrow S} \ G_5 = \emptyset
  }
}{F_3, N, G_3, F_4, (\langle \rangle N \backslash S)/N, G_5 \Rightarrow S/N}
$$

and

$$
\cfrac{
  \cfrac{
    \cfrac{
      \cfrac{\rule{2cm}{0.4pt}}{F_{10}, CN, G_1 \Rightarrow CN} \begin{array}{l} F_{10} = \emptyset \\ G_1 = \emptyset \end{array}
      \quad
      \cfrac{\rule{2cm}{0.4pt}}{F_9, CN, G_8 \Rightarrow CN} \begin{array}{l} F_9 = \emptyset \\ G_8 = \emptyset \\ F_8 = \emptyset \\ F_1 = F_{10} \oplus F_9 \end{array}
    }{F_1, CN, G_1, F_8, CN \backslash CN, G_8 \Rightarrow CN}
  }{F_1, CN, G_1, F_7, []^{-1}(CN \backslash CN), G_7 \Rightarrow CN} \begin{array}{l} F_7 = F_8 \oplus [0] \\ G_7 = [0] \oplus G_8 \end{array}
}{F_1, CN, G_1, F_2, []^{-1}[]^{-1}(CN \backslash CN), G_6 \Rightarrow CN} \begin{array}{l} F_2 = F_7 \oplus [0] \\ G_6 = [0] \oplus G_7 \end{array}
$$

The side-effects for the lowermost application of $/L$, which yields the goal sequent (13), are $G_2 = \emptyset$ and $G_4 = G_5 \oplus G_6$.

Using side effects, the algorithm computes the bracketings from leaves to root as follows:

$$F_{11} = G_9 = F_6 = F_{10} = G_1 = F_9 = G_8 = \emptyset$$
$$F_5 = [0] \oplus F_{11} = [0] \oplus \emptyset = [0]$$
$$G_3 = G_9 \oplus [0] = \emptyset \oplus [0] = [0]$$
$$F_8 = F_4 = G_5 = \emptyset$$
$$F_1 = F_{10} \oplus F_9 = \emptyset \oplus \emptyset = \emptyset$$
$$F_3 = F_6 \oplus F_5 = \emptyset \oplus [0] = [0]$$
$$F_7 = F_8 \oplus [0] = \emptyset \oplus [0] = [0]$$
$$G_7 = [0] \oplus G_8 = [0] \oplus \emptyset = [0]$$
$$F_2 = F_7 \oplus [0] = [0] \oplus [0] = [0,0]$$
$$G_6 = [0] \oplus G_7 = [0] \oplus [0] = [0,0]$$
$$G_2 = \emptyset$$
$$G_4 = G_5 \oplus G_6 = \emptyset \oplus [0,0] = [0,0]$$

and establishes the fact that the following sequent (with ground terms substituted for symbolic bracket variables) is derivable in BI-**Lb**:

$$\emptyset, CN, \emptyset, [0,0], [\,]^{-1}[\,]^{-1}(CN\backslash CN)/(S/N), \emptyset, [0], N, [0], \emptyset, (\langle\rangle N\backslash S)/N, [0,0] \Rightarrow CN.$$

This sequent corresponds to the following **Lb**-sequent:

$$CN, [[[\,]^{-1}[\,]^{-1}(CN\backslash CN)/(S/N), [N], (\langle\rangle N\backslash S)/N]] \Rightarrow CN$$

and the following bracketing of the CN group: "man [[that [Mary] loves]]."

5.3   A Negative Example

In this section we run the bracket-inducing proof search to invalidate (2a), "the paper that John laughed without reading". In order to make the reasoning shorter, we focus on the central part, namely, we show that the dependent clause "John laughed without reading" is not of type S/N. The natural bracketing for this dependent clause would be "[John] laughed [without reading]" (the subject and the without-clause form weak islands), and one can see that the corresponding sequent, according to the lexicon from Fig. 1,

$$[N], \langle\rangle N\backslash S, [[\,]^{-1}((\langle\rangle N\backslash S)\backslash(\langle\rangle N\backslash S))/(\langle\rangle N\backslash S), (\langle\rangle N\backslash S)/N] \Rightarrow S/N$$

is not derivable in **Lb**, since the N which comes from the right-hand side appears outside the bracketed weak island and cannot penetrate the brackets.

Using our bracket-inducing proof search procedure, we establish a stronger fact that there exists *no* bracketing for which the sequent saying that "John laughed without reading" is of type S/N could be derivable in **Lb**.

We shall do proof search in the BI-**Lb** calculus for the following:

(14) $F_1, N, G_1, F_2, \langle\rangle N\backslash S, G_2, F_3, [\,]^{-1}((\langle\rangle N\backslash S)\backslash(\langle\rangle N\backslash S))/(\langle\rangle N\backslash S), G_3,$
$\quad F_4, (\langle\rangle N\backslash S)/N, G_4 \Rightarrow S/N$

and show that it yields no derivation.

The basic idea here is that when the proof search comes to the application of $[]^{-1}L$, there is always going to be $\emptyset$ on the right, thus the side effect for $[]^{-1}L$ would fail, and the pre-derivation will not become a real BI-**Lb** derivation. Accurate justification of this idea requires exhaustive case analysis, which we perform below.

A direct implementation of the algorithm (see Sect. 6) would do a complete proof search. Here we make some optimisations. First we notice that the $/R$ rule is invertible, therefore we can apply it immediately:

$$(15) \quad \begin{array}{c} F_1, N, G_1, F_2, \langle\rangle N\backslash S, G_2, F_3, []^{-1}((\langle\rangle N\backslash S)\backslash(\langle\rangle N\backslash S))/(\langle\rangle N\backslash S), G_3, \\ F_4, (\langle\rangle N\backslash S)/N, G_4, \emptyset, N, \emptyset \Rightarrow S \end{array}$$

Second, we are going to use *count invariants* in order to reduce the number of possible cases to be considered. Count invariants for the Lambek calculus were introduced by van Benthem [2] and then extended to the calculi with brackets and additive connectives [23] and universal exponential [8]. Here we use a very weak form of the count invariant:

(16) **Lemma.** If a sequent is provable in **Lb** (or BI-**Lb**), then each primitive type occurs in it an even number of times.

The proof is a straightforward induction on the derivation of the sequent.

On the top-level, there are three connectives, so we have three cases to consider.

Case 1. Apply $/L$ to the long formula in the center of the sequent (15). Notice that $(\langle\rangle N\backslash S)/N$ and the rightmost N should both go to the left premise, since otherwise it would violate the count invariant and therefore be *a priori* not derivable. The side effect here is $G_3 = \emptyset$, and the right premise is as follows

$$F_1, N, G_1, F_2, \langle\rangle N\backslash S, G_2, F_3, []^{-1}((\langle\rangle N\backslash S)\backslash(\langle\rangle N\backslash S)), \emptyset \Rightarrow S.$$

For this very sequent, further proof search fails, since application of $[]^{-1}L$ is not possible, neither immediately, nor after applying $\backslash L$ to the formula on the left, due to the rightmost $\emptyset$.

Case 2. Apply $/L$ to the rightmost $/$. This yields a side effect $G_4 = \emptyset$ and the (interesting) right premise is as follows:

$$\begin{array}{c} F_1, N, G_1, F_2, \langle\rangle N\backslash S, G_2, F_3, []^{-1}((\langle\rangle N\backslash S)\backslash(\langle\rangle N\backslash S))/(\langle\rangle N\backslash S), \\ G_3, F_4, \langle\rangle N\backslash S, \emptyset \Rightarrow S \end{array}$$

Applying $/L$ to the central formula yields, as the right premise,

$$F_1, N, G_1, F_2, \langle\rangle N\backslash S, G_2, F_3, []^{-1}((\langle\rangle N\backslash S)\backslash(\langle\rangle N\backslash S)), \emptyset \Rightarrow S$$

and again the $\emptyset$ on the right violates the side condition for $[]^{-1}L$, whenever it gets applied.

Applying $\backslash L$ to the right formula makes further derivation impossible, since, due to Lambek's restriction, after that there will be no way to decompose the central formula.

Finally, applying $\backslash L$ to the left formula gives

$$F_1', S, G_2, F_3, []^{-1}(((\langle\rangle N\backslash S)\backslash(\langle\rangle N\backslash S))/(\langle\rangle N\backslash S), G_3, F_4, \langle\rangle N\backslash S, \emptyset \Rightarrow S$$

and we do the same case analysis as above.

Case 3. Apply $\backslash L$ to the leftmost occurrence of $\backslash$. The right premise is as follows:

$$F_1', S, G_2, F_3, []^{-1}(((\langle\rangle N\backslash S)\backslash(\langle\rangle N\backslash S))/(\langle\rangle N\backslash S),$$
$$G_3, F_4, (\langle\rangle N\backslash S)/N, G_4, \emptyset, N, \emptyset \Rightarrow S$$

and we proceed similarly to Cases 1 and 2.

This analysis shows that even if the proof search procedure for (14) successfully finishes at axiom links, resolving side effects fails when it comes across the application of $[]^{-1}L$ due to $\emptyset$ on the right. Therefore, "the paper that John laughed without reading" does not receive type S/N for any bracketing.

6   Complexity Estimations

6.1   Bracket Induction vs. Generate-and-Test Brackets

The proof search algorithm with side-effects presented in the previous section, still has exponential running time. In general, this is inevitable, due to the NP-hardness of the original Lambek calculus [22]. However, the proof search with bracket induction has a significant advantage in speed compared to a naïve approach where the algorithm searches for all possible bracketings by brute force and does proof search independently for each sequent obtained in this way. More precisely, non-determinism in parsing with **Lb**-based categorial grammar comes from three sources:

1. non-unique type assignment (a lexical item can have several different types),
2. bracketing,
3. proof search.

In the bracket induction approach presented in the present paper, the second source above is handled together with the third one. Thus, our algorithm is still exponential, but is also exponentially faster than the naïve one. This makes bracket induction applicable in practice, while attempts to implement brute force bracket guessing fail to parse even simple sentences in reasonable time. An implementation of our parsing algorithm in Prolog, written by the first author, with an example and runtime log showing execution times are available on GitHub: https://github.com/skuzn/BI-Lb

The lower exponential bound on the running time of the bracket-inducing proof search algorithm in the present paper comes from the fact that in the

Lambek calculus, even without brackets, there exist examples of sequents with exponentially many derivations: $P/P, \ldots, P/P, P, P\backslash P, \ldots, P\backslash P \Rightarrow P$. Thus, if we want to yield all possible derivations (parsings) for a sentence, even the output length could be exponential, unless we represent it in a compressed way, as in [5, 21]. The proof search algorithm yields all possible (pre-)derivations in an uncompressed form, and therefore has exponential worst case running time.

6.2 Pseudo-polynomial Approaches

The more subtle question is the comparison of the bracket-inducing parsing procedure presented here with the pseudo-polynomial algorithm for **Lb** presented in our earlier paper [5].

Despite the Lambek calculus being NP-hard [22], Pentus [21] noticed that the complexity essentially comes from complicated types used in the lexicon. He presented a polynomial-time parsing algorithm [21] for Lambek grammars where complexity of all types in the lexicon is bounded. More precisely, the running time of Pentus' algorithm is a polynomial of n and 2^d, where n is the length of the input and d is a complexity measure of types. For simplicity, one can think of d as just the maximal size of a type in the lexicon. Pentus' algorithm is based on proof nets and dynamic programming.

In [5], using the method of Pentus [21], we have presented an algorithm for checking derivability in the Lambek calculus with brackets[2]. However, unlike Pentus' algorithm, our algorithm in [5] is only a theorem-prover, not a parser. That is, the algorithm from [5] does not account for lexical ambiguity, where several types are assigned to one word. Adding this extra level of non-determinism could make running time exponential.

Another, more serious issue is connected with the deep nesting of brackets. The time complexity estimation in [5] is a polynomial of n, 2^d, and n^b, where n is the input length, d is the complexity measure of types in the lexicon, and b is the maximum depth of nested brackets. Thus, the algorithm would run in polynomial time only if both d and b are bound by constants. Unfortunately, in linguistic practice this holds for d, but not for b.

The counter-examples come from well-known phrases with nested dependent clauses, like "the dog that worried the cat that killed the rat that ate the malt that lay in the house that Jack built." The natural bracketing here is as follows (dependent clauses form strong islands, and the subject 'Jack' is a weak one): "the dog [[that worried the cat [[that killed the rat [[that ate the malt [[that lay in the house [[that [Jack] built]]]]]]]]]]." Here b is linear w.r.t. input length ($b = \alpha n$ for some constant α), which yields exponential ($\geq n^{\alpha n}$) running time for the algorithm from [5].

There also exists a shallow bracketing for this phrase: "…the cat [[that killed the rat]] [[that ate the malt]] …" Parsing with this shallow bracketing, however, yields another reading: "the cat ate the malt" rather than the more

---

[2] In contrast to the present paper, the calculus in [5] allows empty antecedents, but imposing the Lambek's restriction there is quite straightforward.

natural "the rat ate the malt." Thus, if we restrict our algorithm by imposing a constant bound on the value of b, we can still justify the phrase as grammatically correct, but we lose some of its readings, which is undesired.

The b parameter being linear w.r.t. n, the algorithm from [5] runs exponentially, as does the algorithm presented in the present paper. The advantage of this latter is that it does not require bracketing to be passed as an input along with the sentence itself.

The question whether there exists an algorithm for **Lb** with the running time being a polynomial of n and 2^d (without n^b in the complexity estimation) is an open problem.

7   Future Work

In order to make the presentation as clear as possible, in this paper we have discussed bracket induction on a very small fragment of type-logical grammar, based on the pure Lambek calculus augmented with brackets and bracket modalities. In the future we are planning to extend this approach to broader calculi, including additive connectives [4], discontinuous operations [16,19], and the (sub)exponential modality for medial and parasitic extraction [17]. For the latter, the whole calculus is undecidable [7], so proof search is possible only in a restricted fragment [7,17]. Moreover, we are planning to optimise parsing with bracket induction using count invariant heuristics [2,8,23] and focusing techniques [1,6,10,18], with necessary modifications for the BI calculi.

Implementing bracket induction in CatLog would allow the system to process raw sentences in natural language, not asking the user for extra structural information (bracketing). Being almost as effective as standard proof search, the proof search procedure with bracket induction would not slow down the parsing workflow. Unfortunately, the running time is still exponential. In the previous section we have discussed why the pseudo-polynomial algorithm for the Lambek calculus with brackets presented in [5] is still not enough to build a polynomial-time version of CatLog. The interesting open question here is whether there is an algorithm for parsing in **Lb** with polynomial runtime for bounded type complexity but unbounded bracket nesting depth, or there is NP-hardness arising from deeply nested brackets even with shallow types.

Acknowledgments. We would like to thank the anonymous referees for their thoughtful comments and questions.

The research of Morrill was supported by the grant TIN2017-89244-R from MINECO (Ministerio de Economia, Industria y Competitividad). Glyn Morrill is also grateful to the University of Pennsylvania for support during his visit in February 2017. Kuznetsov's research towards this paper was supported by the Young Russian Mathematics award, by the Program of the Presidium of the Russian Academy of Sciences No. 01 'Fundamental Mathematics and its Applications' under grant PRAS-18-01, and by the Russian Foundation for Basic Research under grant 18-01-00822. Stepan Kuznetsov is also grateful to the University of Pennsylvania for support during his visit in April–May 2018, when the final version of this paper was prepared. Max

Kanovich is grateful to the University of Pennsylvania for support during his visit in February 2017. The participation of Kanovich, Kuznetsov, and Scedrov in the preparation of this article was within the framework of the Basic Research Program at the National Research University Higher School of Economics (HSE) and supported within the framework of a subsidy by the Russian Academic Excellence Project '5-100'.

References

1. Andreoli, J.M.: Logic programming with focusing in linear logic. J. Log. Comput. **2**(3), 297–347 (1992)
2. van Benthem, J.: Language in Action: Categories, Lambdas, and Dynamic Logic. Studies in Logic and the Foundations of Mathematics, No. 130. North-Holland, Amsterdam (1991). Revised student edition printed in 1995 by the MIT Press
3. Girard, J.Y.: Linear logic. Theor. Comput. Sci. **50**, 1–102 (1987). https://doi.org/10.1016/0304-3975(87)90045-4
4. Kanazawa, M.: The Lambek calculus enriched with additional connectives. J. Log. Lang. Inf. **1**, 141–171 (1992). https://doi.org/10.1007/BF00171695
5. Kanovich, M., Kuznetsov, S., Morrill, G., Scedrov, A.: A polynomial-time algorithm for the Lambek calculus with brackets of bounded order. In: Miller, D. (ed.) Proceedings of the 2nd International Conference on Formal Structures for Computation and Deduction (FSCD 2017). Leibniz International Proceedings in Informatics, LIPIcs, vol. 84, pp. 22:1–22:17. Schloss Dagstuhl - Leibniz-Zentrum für Informatik, Dagstuhl Publishing, Germany (2017). https://doi.org/10.4230/LIPIcs.FSCD.2017.22, http://drops.dagstuhl.de/opus/volltexte/2017/7738/
6. Kanovich, M., Kuznetsov, S., Nigam, V., Scedrov, A.: A logical framework with commutative and non-commutative subexponentials. In: Automated Reasoning: Proceedings of IJCAR 2018. Springer (2018, to appear)
7. Kanovich, M., Kuznetsov, S., Scedrov, A.: Undecidability of the Lambek calculus with subexponential and bracket modalities. In: Klasing, R., Zeitoun, M. (eds.) FCT 2017. LNCS, vol. 10472, pp. 326–340. Springer, Heidelberg (2017). https://doi.org/10.1007/978-3-662-55751-8_26
8. Kuznetsov, S., Morrill, G., Valentín, O.: Count-invariance including exponentials. In: Kanazawa, M., de Groote, P., Sadrzadeh, M. (eds.) Proceedings of the 15th Meeting on the Mathematics of Language, pp. 128–139. Association for Computational Linguistics, London (2017). https://aclweb.org/anthology/W/W17/W17-3413.pdf
9. Lambek, J.: The mathematics of sentence structure. Am. Math. Mon. **65**, 154–170 (1958)
10. Miller, D., Saurin, A.: From proofs to focused proofs: a modular proof of focalization in linear logic. In: Duparc, J., Henzinger, T.A. (eds.) CSL 2007. LNCS, vol. 4646, pp. 405–419. Springer, Heidelberg (2007). https://doi.org/10.1007/978-3-540-74915-8_31
11. Moortgat, M.: Multimodal linguistic inference. J. Log. Lang. Inf. **5**(3, 4), 349–385 (1996). https://doi.org/10.1007/BF00159344
12. Moortgat, M.: Categorial type logics. In: van Benthem, J., ter Meulen, A. (eds.) Handbook of Logic and Language, pp. 93–177. Elsevier Science B.V. and MIT Press, Amsterdam and Cambridge (1997)
13. Morrill, G.: Categorial formalisation of relativisation: pied piping, islands, and extraction sites. Technical report. LSI-92-23-R, Departament de Llenguatges i Sistemes Informàtics, Universitat Politècnica de Catalunya (1992)

14. Morrill, G.: Grammar logicised: relativisation. Linguist. Philos. **40**(2), 119–163 (2017). https://doi.org/10.1007/s10988-016-9197-0

15. Morrill, G.: Parsing logical grammar: CatLog3. In: Loukanova, R., Liefke, K. (eds.) Proceedings of the Workshop on Logic and Algorithms in Computational Linguistics 2017, LACompLing 2017, pp. 107–131. DiVA, Stockholm University (2017). http://su.diva-portal.org/smash/get/diva2:1140018/FULLTEXT03.pdf

16. Morrill, G., Valentín, O.: Displacement calculus. Linguist. Anal. **36**(1–4), 167–192 (2010). arXiv:1004.4181, special issue Festschrift for J. Lambek

17. Morrill, G., Valentín, O.: Computational coverage of TLG: nonlinearity. In: Kanazawa, M., Moss, L., de Paiva, V. (eds.) Proceedings of Third Workshop on Natural Language and Computer Science, NLCS 2015, vol. 32, pp. 51–63. EPiC, Kyoto (2015). Workshop Affiliated with Automata, Languages and Programming (ICALP) and Logic in Computer Science (LICS)

18. Morrill, G., Valentín, O.: Multiplicative-additive focusing for parsing as deduction. In: Cervesato, I., Schürmann, C. (eds.) First International Workshop on Focusing, Workshop Affiliated with LPAR 2015, pp. 29–54. EPTCS, No. 197, Suva, Fiji (2015)

19. Morrill, G., Valentín, O., Fadda, M.: The displacement calculus. J. Log. Lang. Inf. **20**(1), 1–48 (2011). https://doi.org/10.1007/s10849-010-9129-2

20. Morrill, G.V.: Categorial Grammar: Logical Syntax, Semantics, and Processing. Oxford University Press, New York and Oxford (2011)

21. Pentus, M.: A polynomial-time algorithm for Lambek grammars of bounded order. Linguist. Anal. **36**(1–4), 44–471 (2010)

22. Pentus, M.: Lambek calculus is NP-complete. Theor. Comput. Sci. **357**(1), 186–201 (2006). https://doi.org/10.1016/j.tcs.2006.03.018

23. Valentín, O., Serret, D., Morrill, G.: A count invariant for Lambek calculus with additives and bracket modalities. In: Morrill, G., Nederhof, M.-J. (eds.) FG 2012-2013. LNCS, vol. 8036, pp. 263–276. Springer, Heidelberg (2013). https://doi.org/10.1007/978-3-642-39998-5_17

Stripping Isn't so Mysterious, or Anomalous Scope, Either

Daniel Puthawala[⊠]

The Ohio State University, Columbus, USA
`puthawala.1@osu.edu`

Abstract. This paper discusses a common variety of *ellipsis* phenomena
in English called *Stripping*, with particular focus on the observation of so-
called *anomalous scope* of negation and auxiliaries in Stripping sentences,
and the difficulties that this data poses for existing analyses of Stripping.
I then propose an extension to a recent Hybrid Type-Logical Categorical
Grammar account of Gapping that adequately covers Stripping while
straightforwardly accounting for the scope anomalies. This anomalous
scope is a fascinating formal problem on the syntax-semantics interface
that has been thus far overlooked in the stripping literature.

Keywords: Stripping · Anomalous scope · Distributed scope
Hybrid Type-Logical Categorical Grammar · Categorial grammar
Ellipsis

1 Introduction

Stripping is common variety of ellipsis in English. Analyses of this kind of construction, have been stymied by Stripping's uncertain relationship with Gapping,
and the semantic puzzle of anomalous scope. I investigate these problems through
the use of a categorical framework with a flexible Syntax-Semantics interface,
and present an analytical fragment to make steps towards addressing these issues
in tandem.

This paper is organized as follows. Section 2 introduces the main phenomena discussed in this paper, Stripping and Anomalous Scope, in the context of
the problem of Ellipsis and existing work on Gapping. Section 3 demonstrates
why Stripping cannot just be analyzed as a simpler base-case of Gapping, but
is a rather different phenomenon altogether. Section 4 discusses Low-VP Coordination, the main alternative analysis that accounts for Anomalous Scope in
Gapping, and outlines how that analysis could be extended to cover Stripping.
Section 5 outlines the theoretical shortcomings of such an extension, and the
problems that still remain for a satisfactory account of Stripping and Anomalous scope. Section 6 introduces the HTLCG framework as required for the current analytical fragment. Section 7 introduces the analysis of English stripping
in terms of HTLCG. It is also demonstrates how the wide and distributed scope

© Springer-Verlag GmbH Germany, part of Springer Nature 2018
A. Foret et al. (Eds.): FG 2018, LNCS 10950, pp. 102–120, 2018.
https://doi.org/10.1007/978-3-662-57784-4_6

readings of modals and auxiliaries obtain. Finally, Sect. 8 provides a discussion of the larger problem of Anomalous scope in English beyond Stripping and Gapping, and its extensions to other scope-taking modals and auxiliaries than simply negation.

2   Stripping, Gapping, and Anomalous Scope

The problem of ellipsis can be informally defined as cases where the following four facts hold:

(1)   a.   Something is uttered that doesn't look (by itself) to be a "fully formed" sentence.
        b.   The utterance is nonetheless taken to have a "fully-formed" meaning.
        c.   The meaning of the utterance is highly dependent on the context in which it is uttered.
        d.   Speakers, given the same context, generally agree on the exact "fully formed" meaning of the utterance.

It is fairly uncontroversial that (2-a) and (2-b) below are indistinguishable in terms of their truth conditional content.[1] (2-a) is simply an example of sentential coordination. (2-b), on the other hand, is an example of what is commonly referred to in the literature as *Stripping* [4,5,22].

(2)   a.   John ate a burger, and Mary ate a burger (too).
        b.   John ate a burger, and Mary (too).

The following sentences also mean roughly the same thing. Example (3-a) is also again simply sentence coordination, while (3-b) is an example of what is commonly referred to as *Gapping* [7–9,17,18,20].

(3)   a.   John ate a burger, and Mary ate a sub.
        b.   John ate a burger, and Mary a sub.

While (2-b) and (3-b) may appear very similar, a survey of linguistic data reveals that there are generally four main parts of these kinds of constructions, and that these components behave in predictably different ways. In both (2-b) and (3-b), there are two conjuncts. In both cases, the first conjunct, *John ate a burger*, could be a fully satisfactory standalone sentence, while the second conjunct, *Mary (too)* or *Mary a sub*, couldn't, because there is stuff in the first conjunct that is missing from the second one. I refer to this 'missing' material that is only present in the first conjunct as the <u>continuation</u>. The parts of the two conjuncts that aren't missing parallel each other. I call this overt material in the non-sentential conjunct the ASSOCIATE, while I call its counterpart in the more full antecedent conjunct the FOCUS. The bits left over connecting the first

---

[1]   Assuming we aim for the reading of (2-b) in which Mary is an eater, which is a case of subject-stripping, rather than the reading in which John is a cannibal, which would be object-stripping.

and second conjuncts together I shall call the **functor**. Thus in the following Stripping and Gapping examples, the FOCUS and ASSOCIATE are in small caps, the <u>continuation</u> is underlined, and the **functor** is in bold. (4-a) and (4-b) are both examples of Stripping, while (4-c) and (4-d) are parallel Gapping examples.

(4)   a.   **Either** JOHN <u>applied for the job</u>, **or** SANDY.
        b.   <u>Mary told</u> JOHN <u>about the job</u>, **and** SANDY, **(too)**.
        c.   **Either** JOHN <u>applied for</u> THE JOB, **or** SANDY THE GRANT.
        d.   <u>Mary told</u> JOHN <u>about</u> THE JOB, **and** SANDY THE GRANT.

These examples make it plain to see one descriptive generalization that we can use to pretheoretically distinguish between Stripping and Gapping sentences. In the case of Gapping, the FOCUS , "JOHN ... THE JOB" is a non-contiguous string. In the case of Stripping however, the FOCUS , here "JOHN," is a contiguous string. This observation turns out to be a simple and effective way of distinguishing Stripping constructions from Gapping ones without having to appeal to theoretically-motivated assumptions.

One thing that both Stripping and Gapping (among other constructions) exhibit is a peculiar semantic phenomenon whereby negation that appears inside of one disjunct can scope widely over the entire disjunction ([17,21]). Just as bizarre is the observation that in the same sentence, if the reading is forced where negation doesn't have wide scope, it does not have narrow scope just in the first disjunct where it physically appears, but rather is distributed to both disjuncts. Thus, for a sentence such as (5) below, we can obtain readings for wide scope negation as in (5-a), or distributed negation as in (5-b), but the narrow scope negation reading in (5-c) is unavailable. While the distributed reading is expected, the wide scope reading in (5-a) where negation scopes over disjunction is what is referred to as *anomalous scope*. It is important to note that this effect is not limited to negation, however, but occurs for a range of modals and auxiliaries.

(5)   John can't sleep, or Mary.
        a.   Wide-Scope Negation $\neg\diamond(\text{sleep}(j) \vee \text{sleep}(m))$
             (i)   John can't sleep and Mary can't sleep.
        b.   Distributed Negation $\neg(\diamond\text{sleep}(j)) \vee \neg(\diamond\text{sleep}(m))$
             (i)   It's not the case that (both) John can sleep and Mary can sleep.
        c.   Narrow-scope Negation $\neg(\diamond\text{sleep}(j)) \vee \diamond\text{sleep}(m)$
             (i)   It's the case that John can't sleep, or Mary can sleep (or both).

3   Stripping Isn't Just Simple Gapping

Before moving on to analyses of stripping, it is worth taking stock of where we are at the moment, and what the empirical facts of Stripping tell us. Section 2 demonstrated a fairly simple way of telling Stripping and Gapping apart, that being that in the case of Stripping, the FOCUS is a single contiguous string, while the FOCUS in a Gapping sentence can be noncontiguous. It is tempting to thus think of Stripping as simply the simplest possible kind of Gapping, as a sort of contiguous base-case. But even a casual survey of the facts demonstrates that Stripping and Gapping act in very different ways, which reflect the fact that they are surprisingly distinct phenomena, and thus my treatment of Stripping functions very differently than the account of Gapping from which it originates.

One important difference between Stripping and Gapping is that the set of **functors** that can be felicitously used in Stripping constructions only partially overlaps with the set of **functors** in Gapping sentences. This can be seen below where (6)a, c, and e are perfectly fine Stripping constructions, but don't work as Gapping constructions.

(6)   a.   John went to the store before Mary.
      b.   *John went to the store before Mary the beach.
      c.   John went to the store, then Mary.
      d.   *John went to the store, then Mary the beach.
      e.   John went to the store, but not Mary.
      f.   *John went to the store, but not Mary the beach.

In addition, it is widely recognized that Gapping does not allow extraction from embedded clauses. [22, 29] While Stripping is not as free as Pseudogapping, for instance, it can extract from embedded clauses, especially when there is no overt complementizer, such as *that*. Examples such as these also prevent one from being able to treat cases of object stripping as simply NP coordination with unusual prosody.

(7)   a.   John would go to the movies with Linda, but I very much doubt anyone else/Charlie.
      b.   *John would go to the movies with Linda, but I very much doubt Bill Charlie.

The foregoing facts make clear that there are important differences between Stripping and Gapping, which contraindicate any simple assimilation of the former into the latter.

4   Alternative Approach: Stripping as Low-VP Coordination

While to my knowledge there have been no analyses proposed for Stripping that account for the anomalous scope problem described above, such proposals have been made for Gapping. One such line of analysis, first proposed in [7], is the Low VP-Coordination analysis.

It is well accepted in the mainstream generative literature [22, 29] that Stripping is not a movement-based phenomenon like fronting or scrambling, and so most contemporary analyses in that syntactic framework treat Stripping as some sort of coordination and ellipsis. Thus, an alternative analysis to my proposal is to move everything but the remnants out of the conjoined phrase, and either deleting or merging together what's left to obtain the correct surface string. Johnson [7], along with updated versions in [9,10] takes just this tack, and gives the main alternative analysis of Gapping that simultaneously has some success accounting for the scope anomalies of Gapping and Stripping discussed in Sect. 2.

Under Johnson's analysis, what appears to be clausal coordination is actually VP coordination where the subject is removed from the conjuncts via Across-The-Board (ATB) movement, that is symmetrically, and into a position higher in the syntactic tree that scopes over the whole coordinate VP phrase. The second thing that happens is that there is a second special, non-ATB movement whereby the subject of the first conjunct is moved to the Spec position of the matrix AgrP, which preserves the surface word order of Subject1-Verb-VP1-Conj-subject2-VP2, without having the verb trapped within the first conjunct, by similarly allowing the subject of the first conjunct to trivially scope over the entire sentence. His analysis is demonstrated in (8-b) below, where the verb *play* is ATB moved out of both conjuncts to scope over them, and then the subject *Kim* is non-ATB moved to Spec ArgP. This last move is semantically vacuous, and not uncontroversial within Johnson's own research program, but is essential to his analysis of Gapping.

(8) a. Kim didn't play Bingo, or Sandy Chess.
     b. [AgrP Kim$_i$ [Agr didn't [TP play$_j$ [VP [VP [DP ~~Kim$_i$~~] [VP ~~play$_j$~~ Bingo] or [VP [DP Sandy] [VP ~~play$_j$~~ Chess]]]]]]

This analysis is straightforward to adapt to Stripping cases, with two tweaks. First I consider subject Strips, cases where the FOCUS and ASSOCIATE are subjects. For subject Strips, the analysis requires that the entire VP (the continuation) be moved symmetrically out of both conjuncts to a wide scope position instead of just the V node as in (8-b) above.

(9) a. KIM didn't play bingo on Saturday, **or** SANDY.
     b. [AgrP Kim$_i$ [Agr didn't [TP [VP play bingo on Saturday$_j$]] [VP [VP [DP ~~Kim$_i$~~] [VP ~~play bingo on Saturday$_j$~~]] or [VP [DP Sandy] [VP ~~play bingo on Saturday$_j$~~]]]]]]

But Stripping sentences, like Gapping sentences, come in more than one variety. If the FOCUS and ASSOCIATE are objects, rather than subjects, the Low-VP coordination analysis requires a slightly different change. Because now there is asymmetric movement required of the object instead of the subject, we can't simply ATB move the entire VP. However, there is nothing stopping us from ATB moving each part of the VP aside from the Object, as indicated by the traces in (10-b). The three bold movements in (10-b) clearly show how the change of the location of the FOCUS between (9-b) and (10-b) is reflected in the analysis in the form of the asymmetric movement of a DP, while the same component can be ATB moved in the other case.

(10)   a.   Kim didn't play BINGO on Saturday, or CHESS.
        b.   [AgrP **Kim**ᵢ [Agr didn't [TP [VP [VP [V play$_j$] [DP **bingo**$_k$]]
              [PP on Saturday$_l$]] [VP [VP [DP ~~Kim~~ᵢ] [VP [VP [V ~~play~~$_j$] [DP
              ~~bingo~~$_k$]] [PP ~~on Saturday~~$_l$]]] or [VP [DP ~~Kim~~ᵢ] [VP [VP [V
              ~~play~~$_j$] [DP chess]] [PP ~~on Saturday~~$_l$]]]]]]]]

5   Contraindicators to Low-VP Coordination

As we can see, the crux of the Low-VP coordination account is the movement
of what we call the <u>continuation</u> and FOCUS by a mixture of symmetrical ATB
and asymmetrical means to get the word order to work out correctly, thereby
allowing the negative auxiliary to obtain wide scope over the disjunction.

The main problem here is a lack of any independently motivated mechanism
to determine how the multiple elements can be moved and dealt with, while pre-
serving word order. While this problem was noted in [17] for Gapping, it is more
complicated for Stripping. First, this is a problem in the case of a wide-scope
auxiliary such as the modal negation in (10-b), where 'play,' 'bingo,' and 'on
Saturday' must be individually moved out of one or both conjuncts, to somehow
end up back in the same order. This problem only gets worse if we need to get
distributed-scope negation in the semantics, because the negation must originate
in the conjuncts, and then be ATB moved out to its surface position above the
coordinate phrase.

As shown in (11) below, even without the extra complexity introduced by
Johnson's split scope analysis of negative auxiliaries, there are already 24 (4!)
different possible ways that 'didn't,' 'kim,' 'play,' and 'bingo' could be ordered,
But there is no mention of any mechanism for ensuring that these elements end
up in the correct surface ordering.

In addition, given that the movement required in Johnson's account is A'-
movement, there has to be an XP head to move these parts into. But even
in a theory with Larsonian shells, such XPs would be part of the conjuncts
themselves, not the new matrix clause scoping over them. The problem is worse
than just not being able to order the evacuated constituents correctly, there's
no independently-motivated place for them to go! This problem is unchanged
in a movement-and-deletion style analysis where, due to deletion being required
to apply to whole XPs, the idea is to evacuate th surviving constituents out of
the XP, while whatever remains ends up deleted. [6,23] This remains a problem
even there as the initial movement is still problematic.

Thus, even if we were to take solace in the fact that Stripping does not
introduce the vexing problem of a discontinuous FOCUS and ASSOCIATE in the
same way that Gapping does, this analysis would still hit a dead end in precisely
the same manner as described in Kubota and Levine's [17] rebuttal of this kind
of analysis for Gapping. In short, despite the fact that Stripping only allows
a single contiguous FOCUS , this Low-VP Coordination analysis is still equally
unable to account for the required readings of wide and distributed scope.

(11)   K̲i̲m̲ d̲i̲d̲n̲'̲t̲ p̲l̲a̲y̲ BINGO **or** CHESS.
         Distributed reading: ¬(play(bingo)(k)) ∨ ¬(play(chess)(k))

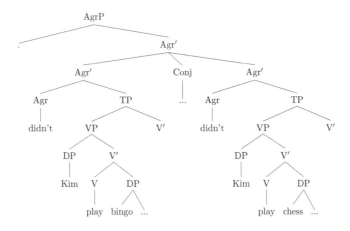

6   Hybrid Type-Logical Categorial Grammar

My analysis of English Stripping and the associated scope anomaly discussed earlier are presented using the Hybrid Type-Logical Categorial Grammar (HTLCG) framework. This section contains the minimum description needed to explain the layout of HTLCG and how it works. The reader is referred to Kubota & Levine, Hybrid Type Logical Categorial Grammar (ms., http://ling.auf.net/lingbuzz/ 002313) for a full formulation of the framework.

HTLCG is based on the Lambek Calculus [19], with one additional nondirectional mode of implication, and developed in Kubota [11,15], and Kubota and Levine [12–14,16,17]. The flexible syntax-semantics interface of this system is useful for studying linguistic phenomena that have implications in both domains, such as coordination, scope, and ellipsis. Readers familiar with HTLCG may wish to skip straight to Sect. 7.1.

Linguistics expressions in HTLCG are represented as tuples $\langle \phi; \sigma; \kappa \rangle$ where ϕ is the phonological string, σ is the semantic term, and κ is the syntactic type. In this framework, syntactic *type* is synonymous with syntactic *category*. Examples (12-a,b) below are NPs, (12-c) is a one-place predicate, and (12-d) is a two-place predicate.

(12)   Sample Lexicon:
         a. John; j; NP
         b. bingo; b; NP
         c. sleeps; sleep; NP\S
         d. eats; eat; (NP\S)/NP

The / and \ are the normal mode of Lambek directional implication where the argument falls under the slash and the functor is above. Thus A/B is an expression looking for an argument of type B on its right, while A\B denotes an expression looking for an argument of type A on it's left. In HTLCG there is one additional mode of implication, $\upharpoonright$, which is a directionless mode of implication similar to $\multimap$ in Curryesque categorial grammars. While both / and \ are directional modes of implication that are looking for adjacent peripheral arguments, $\upharpoonright$ is able to take arguments from anywhere, and is often used to denote continuations, an expression missing an argument from a medial position within itself, and other discontinuous linguistic expressions.

(13)   HTLCG Elimination rules

$$\frac{\text{a; F; A/B} \qquad \text{b; G; B}}{\text{a·b; F(G); A}} /E \qquad \frac{\text{b; G; B} \qquad \text{a; F; B\backslash A}}{\text{b·a; F(G); A}} \backslash E$$

$$\frac{\text{a; F; A}\upharpoonright\text{B} \qquad \text{b; G; B}}{\text{a(b); F(G); A}} \upharpoonright E$$

The elimination rules are different modes of modus ponens. The /E rule allows a functor to take an argument on its right periphery, while the \E rule allows the same but on the left periphery. the $\upharpoonright$E rule allows the functor to take an appropriately typed argument from anywhere. While this may seem overly powerful, phonological string ordering is still maintained via functional prosody.[2] The · connective is for phonological string concatenation, and is associative in both directions.

---

[2] One reviewer points out a potential problem with the underlying logic of the $\upharpoonright$ connective. As discussed in [25], it can allow for undesired overgeneration, particularly in cases such as determiner gapping and stripping. The problem is that the $\upharpoonright$E rule only requires that the syntactic categories match, and is insensitive to the end linear order resulting from prosodic function-application. This means that it does not necessarily require the end result of prosodic function-application to match the order of the hypothetical expressions used in the $\upharpoonright$I rule to derive the original continuation in the first place.

However, the author does not consider this criticism to be an existential threat to the present analysis for several reasons. First, while it is clear that the current formulation of the $\upharpoonright$, coupled with its use in some lexical entries, is problematic, further research is required to determine if this issue can be solved through minor tweaks to the system or if it will require wholesale revisions of the underlying logic. Secondly, the present analysis, though couched HTLCG, is readily adaptable into other CG and TLG frameworks, such as the Displacement Type-Logical Grammar of [24], as noted by Morrill and Valentin in [25]. Thus, even if this observation proves a major obstacle for HTLCG as a framework in its current form, it would not necessarily invalidate the results of the current analysis.

(14)   HTLCG Introduction Rules

$$\frac{\begin{array}{c}[\phi;\ \text{x};\ \text{A}]^n \\ \vdots \quad \vdots \quad \vdots \\ \dfrac{\text{b}\cdot\phi;\ \text{F};\ \text{B}}{\text{b};\ \lambda\text{x}[\text{F}];\ \text{B/A}}\end{array}}{}\ /\text{I}^n \qquad \frac{\begin{array}{c}[\phi;\ \text{x};\ \text{A}]^n \\ \vdots \quad \vdots \quad \vdots \\ \dfrac{\phi\cdot\text{b};\ \text{F};\ \text{B}}{\text{b};\ \lambda\text{x}[\text{F}];\ \text{A}\backslash\text{B}}\end{array}}{}\ \backslash\text{I}^n$$

$$\frac{\begin{array}{c}[\phi;\ \text{x};\ \text{A}]^n \\ \vdots \quad \vdots \quad \vdots \\ \text{b};\ \text{F};\ \text{B}\end{array}}{\lambda\phi[\text{b}];\ \lambda\text{x}[\text{F}];\ \text{B}{\upharpoonright}\text{A}}\ {\upharpoonright}\text{I}^n$$

The introduction rules enable hypothetical reasoning, and are typically considered more abstract and harder to intuit, but play an equally important role in the overall logic of the deductive system. For the purposes of the present analysis, the introduction rules are often used when type-raising and lowering linguistic expressions, and when deriving the signs corresponding to noncontiguous constituents, such as the aptly named <u>continuation</u> from (4-b) above, *Mary told...about the job.* To derive that expression, we could hypothesize an NP, for example, $[\phi;\ \text{x};\ \text{NP}]^i$, where the superscript i is tracking the index of the hypothesized element. After using the elimination rules to derive the string *Mary·told·ϕ·about·the·job* of type S, we could use the ${\upharpoonright}$I rule to discharge our hypothesized NP and obtain the type of our continuation in (15), which denotes a discontiguous expression that would be an S, if it had a medial NP argument.

(15) $\lambda\phi[\text{Mary·told·}\phi\text{·about·the·job}];\ \lambda\text{x}[\text{told}(\text{about job})(\text{x})(\text{m})];\ \text{S}{\upharpoonright}\text{NP}$

There is a transparent syntax-semantics mapping from syntactic categories to semantic types in HTLCG. As issues related to the intensionality or extensionality of our semantics are not directly related to the problem of stripping, this analysis employs a fragment of standard extensional Montagovian model-theoretic semantics in the current analysis as described in Kubota and Levine [2014] and [2015a]:

(16)   a. e and t are semantics types
        b. if α and β are semantic types, then so is $\alpha \to \beta$
        c. Nothing else is a semantic type

Syntactic categories are can be mapped to semantic types by a function SEM, defined below.

(17)   a. $\text{SEM}(\text{NP}) = e$
        b. $\text{SEM}(\text{S}) = t$
        c. $\text{SEM}(\text{S}_{\text{bse}}) = t$
        d. $\text{SEM}(\mathbb{W}) = t$
        e. For any categories A and B:

$$(i).\ \text{SEM}(A/B) = \text{SEM}(B) \to \text{SEM}(A)$$
$$(ii).\ \text{SEM}(A \backslash B) = \text{SEM}(A) \to \text{SEM}(B)$$
$$(iii).\ \text{SEM}(A{\restriction}B) = \text{SEM}(B) \to \text{SEM}(A)$$

In addition, for the phonological string, we can define a function SAY to map the syntactic category to a phonological type.

(18) a. For any atomic syntactic type A:
         (i) SAY(A) = *string*
      b. For any categories A and B:
         (i) SAY(A/B) = *string*
         (ii) SAY(A\B) = *string*
         (iii) SAY(A↾B) = SAY(B)→ SAY(A)

In the sections and analytical fragment that follows, I make use of the following variables in (19).

(19) a. Prosodic variables:
         (i) ϕ, ψ - *string*
         (ii) σ - *string* → *string*
      b. Semantic Variables:
         (i) x, y, z - e
         (ii) P, Q - $e \to t$
         (iii) R - $(e \to t) \to (e \to t)$

In addition to the normal prosodic and semantic variables, HTLCG also includes metavariable over syntactic categories, as a kind of type schema. This allows HTLCG to capture certain empirical facts about language more efficiently by generalizing lexical entries. For instance, the category of conjunction in English is typically taken to be X\X/X. That is, something that, for any category X, returns an expression of type X if there is an expression of type X on its right and its left. Importantly, however, when such a lexical entry is introduced as part of a proof, all instances of the same metavariable receive the same category assignment. Thus the X\X/X entry for conjunction above could be realized in a proof as NP\NP/NP, S\S/S, or even (NP\N)\(NP\N)/(NP\N). However, NP\(NP\N)/(NP\N) would not be a legal type assignment, because it would require X to simultaneously be of type NP in one instance, and type NP\S in another.

We will use the following primitive syntactic categories for this fragment. NP is the syntactic category of Noun Phrases such as *John* and *bingo*. S is the syntactic category of tensed clauses, which are acceptable final outputs of the grammar. S$_{\text{bse}}$ is the type of base or untensed clauses. Finally we have $\mathbb{W}$, a so-called "poltergeist category." $\mathbb{W}$ is a type that can be legally used in the course of a derivation, but there are no lexical constants of type $\mathbb{W}$, so it a useful, albeit uninhabited type, in the spirit of [26], though the actual implementation here is different from the original version in GPSP. $\mathbb{W}$ is a category that corresponds to Troelstra's 0 in intuitionist linear logic, representing multiplicative False in the syntax in HTLCG. [28] The grammar does not generate any signs of just category $\mathbb{W}$ in the lexicon, and the tecto logic of categories cannot prove $\mathbb{W}$. In effect, a derivation that results in a $\mathbb{W}$ indicates ungrammaticality.

7 HTLCG Analysis of Stripping

7.1 Basic Stripping

The following proof demonstrates how compositional meaning of a very simple Stripping sentence, *John slept, and Mary too* can be derived, and introduces the Stripping operator in (21) below. First the ASSOCIATE and FOCUS are fed to the Stripping operator, which in its current version is baked into the MAIN FUNCTOR. Then we feed the <u>continuation</u> to the **functor**, and β-reduce to distribute the semantics of the <u>continuation</u> to the two conjuncts.

In the phonological tuple, the Stripping operator in (20-d) takes in the ASSOCIATE before taking in the FOCUS and passing that string along to the <u>continuation</u>. In the semantics, the operator similarly takes both the ASSOCIATE and FOCUS, but in this case it passes both arguments along to the <u>continuation</u>, *slept*, in the final step of the derivation. In this way the operator obtains the desired surface string and semantic denotation of Stripping by distributing the meaning of the <u>continuation</u> to both conjuncts, without a corresponding symmetric appearance of the phonological string.

(20)   John slept, and Mary, too.
   a. John; j; NP
   b. Mary; m; NP
   c. slept; sleep; NP\S from which we can freely derive [3]:
      (i) $\lambda\phi[\phi\cdot\text{slept}]$; $\lambda y[\text{sleep}(y)]$; S↑NP
   d. $\lambda\phi\lambda\psi\lambda\sigma[\sigma(\psi)\cdot\text{and}\cdot\phi\cdot\text{too}]$; $\lambda y\lambda x\lambda P[P(x)\wedge P(y)]$; Y↑(Y↑X)↑X↑X

(21)

$$
\cfrac{\begin{array}{c}\text{John;}\\ \text{j;}\\ \text{NP}\end{array} \quad \cfrac{\cfrac{\begin{array}{c}\lambda\phi\lambda\psi\lambda\sigma[\sigma(\psi)\cdot\text{and}\cdot\phi\cdot\text{too}];\\ \lambda y\lambda x\lambda P[P(x)\wedge P(y)];\\ \text{S↑(S↑NP)↑NP↑NP}\end{array} \quad \begin{array}{c}\text{Mary;}\\ \text{m;}\\ \text{NP}\end{array}}{\begin{array}{c}\lambda\psi\lambda\sigma[\sigma(\psi)\cdot\text{and}\cdot\text{Mary}\cdot\text{too}];\\ \lambda x\lambda P[P(x)\wedge P(m)];\\ \text{S↑(S↑NP)↑NP}\end{array}} \scriptstyle{\uparrow E}}{\begin{array}{c}\lambda\sigma\cdot[\sigma(\text{John})\cdot\text{and}\cdot\text{Mary}\cdot\text{too}];\\ \lambda P[P(j)\wedge P(m)];\\ \text{S↑(S↑NP)}\end{array}} \scriptstyle{\uparrow E} \quad \begin{array}{c}\lambda\phi[\phi\cdot\text{slept}];\\ \lambda y[\text{sleep}(y)];\\ \text{S↑NP}\end{array}}{\text{John}\cdot\text{slept}\cdot\text{and}\cdot\text{Mary}\cdot\text{too; sleep(j)}\wedge\text{sleep(m); S}} \scriptstyle{\uparrow E}
$$

We now turn to negation Stripping, as in (22) below. As in the case of conjunction Stripping, this kind of Stripping sentence can also be compositionally derived in a straightforward manner with only minor changes to the phonological and semantic tuples of the Stripping operator's lexical entry. Similarly to the conjunction Stripping derivation, we first feed the ASSOCIATE and FOCUS to the operator built into the MAIN FUNCTOR, before feeding in the CONTINUATION, and then β-reducing to obtain our final sentence.

---

[3] This vertically-slashed version of *slept* can be derived simply via hypothetical reasoning:

$$
\cfrac{[\phi;\ y;\ \text{NP}]^i \quad \text{slept; sleep; NP\S}}{\cfrac{\phi\cdot\text{slept; sleep(y); S}}{\lambda\phi[\phi\cdot\text{slept}];\ \lambda y[\text{sleep}(y)];\ \text{S↑NP}}\scriptstyle{\uparrow I^i}}
$$

(22)   John slept, not Mary.

a. $\lambda\phi\lambda\psi\lambda\sigma[\sigma(\psi)\cdot\text{not}\cdot\phi]$; $\lambda y\lambda x\lambda P[P(x)\wedge\neg P(y)]$; $Y{\upharpoonright}(Y{\upharpoonright}X){\upharpoonright}X{\upharpoonright}X$

$$
\frac{
\begin{array}{c}
\dfrac{
\begin{array}{c}
\dfrac{
\begin{array}{c}
\lambda\phi\lambda\psi\lambda\sigma[\sigma(\psi)\cdot\text{not}\cdot\phi]; \\
\lambda y\lambda x\lambda P[P(x)\wedge\neg P(y)]; \\
S{\upharpoonright}(S{\upharpoonright}NP){\upharpoonright}NP{\upharpoonright}NP
\end{array}
\quad \text{Mary; m; NP}
}{
\begin{array}{c}
\lambda\psi\lambda\sigma[\sigma(\psi)\cdot\text{not}\cdot\text{Mary}]; \\
\lambda x\lambda P[P(x)\wedge\neg P(m)]; \\
S{\upharpoonright}(S{\upharpoonright}NP){\upharpoonright}NP
\end{array}
}\ {\upharpoonright}E
}{\lambda\sigma\cdot[\sigma(\text{John})\cdot\text{not}\cdot\text{Mary}];\ \lambda P[P(j)\wedge\neg P(m)];\ S{\upharpoonright}(S{\upharpoonright}NP)}\ {\upharpoonright}E
\end{array}
\quad
\begin{array}{c}
\lambda\phi[\phi\cdot\text{slept}]; \\
\lambda y[\text{sleep}(y)]; \\
S{\upharpoonright}NP
\end{array}
}{\text{John}\cdot\text{slept}\cdot\text{not}\cdot\text{Mary}];\ \text{sleep}(j)\wedge\neg\text{sleep}(m);\ S}\ {\upharpoonright}E
$$

where "John; j; NP" appears to the left.

Similarly to the two above examples, other kinds of basic Stripping constructions such as those below in (23) can be derived with only minor tweaks to the Stripping operator.

(23)   a. John slept, and Mary.
        b. John slept, or Mary.
        c. Either John slept, or Mary.
        d. Neither John slept, nor Mary.
        e. John slept, then Mary.

This analysis for stripping also obtains for object and medial stripping, as demonstrated below for medial object stripping, which cannot otherwise be straightforwardly captured by a coordination analysis. First, we derive the <u>continuation</u>, Mary edited ϕ in the park. Then we feed the ASSOCIATE and FOCUS to the Stripping operator as before, and finally combine to derive an S.

(24)   Mary edited the paper in the park, and the journal, too.

(25)

$$
\frac{
\begin{array}{c}
\text{Mary;} \\ \text{m;} \\ \text{NP}
\end{array}
\quad
\dfrac{
\dfrac{
\dfrac{
\begin{array}{c}\text{edited; edit;} \\ (NP\backslash S)/NP\end{array}\quad [\phi;\ x;\ NP]^i
}{\begin{array}{c}\text{edited}\cdot\phi; \\ \text{edit}(x);\ NP\backslash S\end{array}}\ /E
\quad
\begin{array}{c}\text{in}\cdot\text{the}\cdot\text{park; inpark;} \\ (NP\backslash S)\backslash(NP\backslash S)\end{array}
}{\begin{array}{c}\text{edited}\cdot\phi\cdot\text{in}\cdot\text{the}\cdot\text{park;} \\ \text{inpark}(\text{edit}(x));\ NP\backslash S\end{array}}\ \backslash E
}{}
}{\text{Mary}\cdot\text{edited}\cdot\phi\cdot\text{in}\cdot\text{the}\cdot\text{park; inpark}(\text{edit}(x)(m));\ S}\ \backslash E
$$
$$
\frac{}{\lambda\phi[\text{Mary}\cdot\text{edited}\cdot\phi\cdot\text{in}\cdot\text{the}\cdot\text{park}];\ \lambda x[\text{inpark}(\text{edit}(x)(m))];\ S{\upharpoonright}NP}\ {\upharpoonright}I^i
$$

(26)

$$
\frac{
\begin{array}{c}\text{the}\cdot\text{paper;} \\ \text{paper;} \\ NP\end{array}
\quad
\dfrac{
\dfrac{
\begin{array}{c}
\lambda\phi\lambda\psi\lambda\sigma[\sigma(\psi)\cdot\text{and}\cdot\phi\cdot\text{too}]; \\
\lambda y\lambda x\lambda P[P(x)\wedge P(y)]; \\
S{\upharpoonright}(S{\upharpoonright}NP){\upharpoonright}NP{\upharpoonright}NP
\end{array}
\quad
\begin{array}{c}\text{the}\cdot\text{journal;} \\ \text{journal;} \\ NP\end{array}
}{
\begin{array}{c}
\lambda\psi\lambda\sigma[\sigma(\psi)\cdot\text{and}\cdot\text{the}\cdot\text{journal}\cdot\text{too}]; \\
\lambda x\lambda P[P(x)\wedge P(\text{journal})]; \\
S{\upharpoonright}(S{\upharpoonright}NP){\upharpoonright}NP
\end{array}
}\ {\upharpoonright}E
}{}
}{
\begin{array}{c}
\lambda\sigma[\sigma(\text{the}\cdot\text{paper})\cdot\text{and}\cdot\text{the}\cdot\text{journal}\cdot\text{too}]; \\
\lambda P[P(\text{paper})\wedge P(\text{journal})]; \\
S{\upharpoonright}(S{\upharpoonright}NP)
\end{array}
}\ {\upharpoonright}E
$$

$$(27) \quad \frac{\begin{array}{cc} \lambda\phi[\text{Mary·edited·}\phi\text{·in·the·park}]; & \lambda\sigma[\sigma(\text{the·paper})\text{·and·the·journal·too}]; \\ \lambda x[\text{inpark}(\text{edit}(x)(m))]; & \lambda P[P(\text{paper})\wedge P(\text{journal})]; \\ S\uparrow NP & S\uparrow(S\uparrow NP) \end{array}}{\begin{array}{c} \text{Mary·edited·the·paper·in·the·park·and·the·journal·too;} \\ \text{inpark}(\text{edit}(\text{paper})(m))\wedge\text{inpark}(\text{edit}(\text{journal})(m)); \\ S \end{array}}$$

7.2 Wide Scope of Negation

This section briefly presents derivations of the following sentences in (28) to build up to a proper analysis of Stripping sentences with wide scope negation under disjunction.

Derivations are provided below for the following examples in order of increasing complexity.

(28)   a. John didn't sleep.
        b. John didn't sleep, or Mary.

We then consider wide-scope negation under disjunction demonstrating an approach to negation, which is applicable more generally than in (22) above, where negation was built into the **functor**. To this end, we adopt an analysis of negation similar to the prior analyses of modal auxiliaries as VP/VP operators along the lines of [17], and [27], in the tradition of Bach [1–3], yielding a straightforward integration of modals into the current analysis. To demonstrate this consider the derivation below of the sentence in (28-a).

(29)   Lexical entry for Auxiliaries:
        $\lambda\sigma[\sigma(\text{didn't})]; \lambda\mathfrak{F}[\neg\mathfrak{F}(\lambda P[P])]; S\uparrow(\mathbb{W}\uparrow((NP\backslash\mathbb{W})/(NP\backslash S_{\text{bse}})))$

Example (29) above is our lexical entry for the auxiliary *didn't*. As mentioned previously, if one is not concerned with agreement, it's possible for the purposes of this analysis to think of $\mathbb{W}$ as a funny looking S_{bse}, in which case the syntactic type of the above auxiliary resembles the more familiar $S\uparrow((S_{\text{bse}}\uparrow(NP\backslash S_{\text{bse}})/(NP\backslash S_{\text{bse}})))$, an expression that returns an S if given something that would be an S_{bse} but is missing a $(NP\backslash S_{\text{bse}})/(NP\backslash S_{\text{bse}})$ modifier in some medial position. In the proof of (28-a) below, we can see how this auxiliary can be used in a normal sentence without stripping or other ellipsis. We first posit ϕ, a hypothetical Verb Phrase modifier, and combine it with *sleep* and *John*, before we then discharge that hypothesis via $\uparrow$I to obtain a continuation $\lambda\phi[\text{John·}\phi\text{·sleep}]$. This expression is then taken as an argument by the higher order auxiliary. An alternative proof strategy is, rather than type-raising *John... sleep*, to instead lower *didn't* into a lower-order (ie Lambek-slashed) category as demonstrated later on in (36).

(30)   Derivation of (28-A) *John didn't sleep.*

$$
\cfrac{
\lambda\sigma[\sigma(\text{didn't})];\\
\lambda\mathfrak{F}[\neg\mathfrak{F}(\lambda P[P])];\\
S{\upharpoonright}(\mathbb{W}{\upharpoonright}((NP\backslash\mathbb{W})/(NP\backslash S_{bse})))
\quad
\cfrac{
\cfrac{
\text{John};\\
\text{j};\\
NP
\quad
\cfrac{
\begin{array}{cc}[\phi;&\text{sleep};\\R;&\text{sleep};\\(NP\backslash\mathbb{W})/(NP\backslash S_{bse})]^a&(NP\backslash S_{bse})\end{array}
}{\phi{\cdot}\text{sleep};\ R(\text{sleep});\ (NP\backslash\mathbb{W})}/E
}{\text{John}{\cdot}\phi{\cdot}\text{sleep};\ R(\text{sleep}(\text{j}));\ \mathbb{W}}\backslash E
}{\lambda\phi[\text{John}{\cdot}\phi{\cdot}\text{sleep}];\ \lambda R[R(\text{sleep}(\text{j}))];\ \mathbb{W}{\upharpoonright}((NP\backslash\mathbb{W})/(NP\backslash S_{bse}))}{\upharpoonright}I^a
}{\text{John}{\cdot}\text{didn't}{\cdot}\text{sleep};\ \neg(\text{sleep}(\text{j}));\ S}{\upharpoonright}E
$$

All necessary elements are now available to derive a proof of (28-b) *John didn't sleep or Mary* using a disjunction version of the Stripping operator from Sect. 1 above:

(31) $\lambda\phi\lambda\psi\lambda\sigma[\sigma(\psi){\cdot}\text{or}{\cdot}\phi];\ \lambda y\lambda x\lambda P[P(x)\vee P(y)];\ Y{\upharpoonright}(Y{\upharpoonright}X){\upharpoonright}X{\upharpoonright}X$

Since we are Stripping again now, we need a vertically-slashed type predicate with the appropriate arguments to be saturated later on via hypothetical reasoning.

(32)
$$
\cfrac{
\begin{array}{c}[\psi;\\x;\\NP]^b\end{array}
\quad
\cfrac{
\begin{array}{cc}[\phi;&\text{sleep};\\R;&\text{sleep};\\(NP\backslash\mathbb{W})/(NP\backslash S_{bse})]^a&(NP\backslash S_{bse})\end{array}
}{\phi{\cdot}\text{sleep};\ R(\text{sleep});\ (NP\backslash\mathbb{W})}/E
}{
\cfrac{\psi{\cdot}\phi{\cdot}\text{sleep};\ R(\text{sleep})(\psi);\ \mathbb{W}}{\lambda\psi[\psi{\cdot}\phi{\cdot}\text{sleep}];\ \lambda x[R(\text{sleep}(x))];\ \mathbb{W}{\upharpoonright}NP}{\upharpoonright}I^b}
\backslash E
$$

We supply signs corresponding to the FOCUS and ASSOCIATE to the Stripping operator, and then pick up the $\mathbb{W}{\upharpoonright}NP$-typed predicate to yield *John ϕ sleep or Mary.* We then need to discharge the other assumed $(NP\backslash\mathbb{W})/(NP\backslash S_{bse})$ expression to obtain a syntactic type that *didn't* is looking for.

(33)
$$
\cfrac{
\begin{array}{c}\lambda\psi[\psi{\cdot}\phi{\cdot}\text{sleep}];\\\lambda x[R(\text{sleep}(x))];\\\mathbb{W}{\upharpoonright}NP\end{array}
\quad
\cfrac{
\cfrac{
\begin{array}{c}\text{John};\\\text{j};\\NP\end{array}
\quad
\cfrac{
\begin{array}{c}\text{Mary};\\\text{m};\\NP\end{array}
\quad
\begin{array}{c}\lambda\phi\lambda\psi\lambda\sigma[\sigma(\psi){\cdot}\text{or}{\cdot}\phi];\\\lambda y\lambda x\lambda P[P(x)\vee P(y)];\\\mathbb{W}{\upharpoonright}(\mathbb{W}{\upharpoonright}NP){\upharpoonright}NP{\upharpoonright}NP\end{array}
}{\begin{array}{c}\lambda\psi\lambda\sigma[\sigma(\psi){\cdot}\text{or}{\cdot}\text{Mary}];\\\lambda x\lambda P[P(x)\vee P(m)];\\\mathbb{W}{\upharpoonright}(\mathbb{W}{\upharpoonright}NP){\upharpoonright}NP\end{array}}{\upharpoonright}E
}{\begin{array}{c}\lambda\sigma[\sigma(\text{John}){\cdot}\text{or}{\cdot}\text{Mary}];\\\lambda P[P(\text{j})\vee P(m)];\\\mathbb{W}{\upharpoonright}(\mathbb{W}{\upharpoonright}NP)\end{array}}{\upharpoonright}E
}{
\cfrac{\text{John}{\cdot}\phi{\cdot}\text{sleep}{\cdot}\text{or}{\cdot}\text{Mary};\ R(\text{sleep}(\text{j}))\vee R(\text{sleep}(m))\ \mathbb{W}}{\lambda\phi[\text{John}{\cdot}\phi{\cdot}\text{sleep}{\cdot}\text{or}{\cdot}\text{Mary}];\ \lambda R[R(\text{sleep}(\text{j}))\vee R(\text{sleep}(m))];\ \mathbb{W}{\upharpoonright}((NP\backslash\mathbb{W})/(NP\backslash S_{bse}))}{\upharpoonright}I^a}
$$

The VP operator *didn't* then combines with the sign derived in (33), yielding the correct semantics for (28-b).

(34)
$$
\cfrac{
\begin{array}{c}\lambda\sigma[\sigma(\text{didn't})];\\\lambda\mathfrak{F}[\neg\mathfrak{F}(\lambda P[P])];\\S{\upharpoonright}(\mathbb{W}{\upharpoonright}((NP\backslash\mathbb{W})/(NP\backslash S_{bse})))\end{array}
\quad
\begin{array}{c}\lambda\phi[\text{John}{\cdot}\phi{\cdot}\text{sleep}{\cdot}\text{or}{\cdot}\text{Mary}];\\\lambda R[R(\text{sleep})(\text{j})\vee R(\text{sleep})(m)];\\\mathbb{W}{\upharpoonright}((NP\backslash\mathbb{W})/(NP\backslash S_{bse}))\end{array}
}{\begin{array}{c}\text{John}{\cdot}\text{didn't}{\cdot}\text{sleep}{\cdot}\text{or}{\cdot}\text{Mary};\\\neg(\text{sleep}(\text{j})\vee\text{sleep}(m));\\S\end{array}}
$$

Finally, suppose we wanted to derive the other version of (28-b), but with the distributed-scope negation reading, as in (35-b) below:

(35)   a. John didn't sleep or Mary.
       b. $\neg$sleep(j) $\vee$ $\neg$sleep(m)

This is not a problem for my analysis, and falls out naturally from the lexicon and fragment presented thus far. We are simply required to derive a lower-order version of the auxiliary.

Deriving Lower-Order Auxiliaries. In HTLCG, we can derive a lower-order version of our (by default) higher-order VP/VP auxiliaries by hypothesizing the needed arguments, and then discharging those hypotheses on the periphery. Kubota and Levine [2016] contains an identical derivation in their appendices, demonstrating how to lower the modal auxiliary. I repeat it here for convenience.

HTLCG assumes that modal auxiliaries are higher-order Verb Phrase modifiers. As demonstrated in Sect. 6.2, this higher-order entry yields the wide-scope modal and negation semantics and partially solves the problem of anomalous scope. But oftentimes a lower-order version of these same modals are required, such as in cases of distributed scope. However, this does not require HTLCG to have two different entries for auxiliaries. Rather, the lower-order version can be derived as a theorem from the higher-order entry via hypothetical reasoning.

In the place of *john* and *sleep* in (36), we posit variables to derive an expression of type $\mathbb{W}\!\upharpoonright\!((NP\backslash\mathbb{W})/(NP\backslash S_{bse}))$ which then combines with our higher-order auxiliary to yield an S. Then, by directional slash elimination, we derive an expression of type NP\S. finally, with one more directional slash introduction, we obtain $(NP\backslash S)/(NP\backslash S_{bse})$, the type of lower-order auxiliary.

(36)

$$
\cfrac{\cfrac{\lambda\sigma[\sigma(didn't)];\ \lambda\mathfrak{F}[\neg\mathfrak{F}(\lambda P[P])];\ S\!\upharpoonright\!(\mathbb{W}\!\upharpoonright\!((NP\backslash\mathbb{W})/(NP\backslash S_{bse})))\quad \cfrac{\cfrac{[\phi;\ x;\ NP]^a\quad \cfrac{[\phi;\ R;\ (NP\backslash\mathbb{W})/(NP\backslash S_{bse})]^b\quad [\phi;\ Q;\ (NP\backslash S_{bse})]^c}{\phi^b\!\cdot\!\phi^c;\ R(Q);\ (NP\backslash\mathbb{W})}}{\phi^a\!\cdot\!\phi^b\!\cdot\!\phi^c;\ R(Q)(x);\ \mathbb{W}}}{\lambda\phi^b[\phi^a\!\cdot\!\phi^b\!\cdot\!\phi^c];\ \lambda R[R(Q)(x)];\ \mathbb{W}\!\upharpoonright\!((NP\backslash\mathbb{W})/(NP\backslash S_{bse}))}\ \upharpoonright I^b}{\cfrac{\cfrac{\cfrac{\lambda\phi^b[\phi^a\!\cdot\!\phi^b\!\cdot\!\phi^c]\ (didn't);\ \neg\lambda R[R(Q)(x)](\lambda P[P]);\ S}{\phi^a\!\cdot\!didn't\!\cdot\!\phi^c;\ \neg\lambda P[P](Q)(x);\ S}\ \beta\Rightarrow}{\phi^a\!\cdot\!didn't\!\cdot\!\phi^c;\ \neg Q(x);\ S}\ \beta\Rightarrow}{\cfrac{didn't\!\cdot\!\phi^c;\ \lambda x[\neg Q(x)];\ NP\backslash S}{didn't;\ \lambda Q\lambda x[\neg Q(x)];\ (NP\backslash S)/(NP\backslash S_{bse})}\ /I^c}\ \backslash I^a}
$$

With the lower-order auxiliary in hand as a theorem of the higher-order lexical entry and the inference rules of HTLCG, we straightforwardly obtain the distributed reading of negation and modals in Stripping.

First we once again derive the vertically-slashed type of *sleep*, but this time our hypothetical auxiliary is $(NP\backslash S)/(NP\backslash S_{bse})$ rather than $(NP\backslash\mathbb{W})/(NP\backslash S_{bse})$:

(37)
$$\cfrac{\cfrac{[\psi;\ x;\ NP]^b \qquad \cfrac{[\phi;\ R;\ (NP\backslash S)/(NP\backslash S_{bse})]^a \qquad sleep;\ sleep;\ (NP\backslash S_{bse})}{\phi\cdot sleep;\ R(sleep);\ NP\backslash S}}{\psi\cdot\phi\cdot sleep;\ R(sleep)(x);\ S}}{\lambda\psi[\psi\cdot\phi\cdot sleep];\ \lambda x[R(sleep)(x)];\ S\!\upharpoonright\!NP}\ \upharpoonleft I^b$$

We then feed our FOCUS, ASSOCIATE, and verb to the Stripping operator, and β-reduce, exactly the same as in (33) above.

(38)
$$\cfrac{\cfrac{\cfrac{\lambda\psi[\psi\cdot\phi\cdot sleep];}{\lambda x[R(sleep)(x)];}\quad John;\ j;\ NP \quad \cfrac{Mary,\ m,\ NP \quad \cfrac{\lambda\phi\lambda\psi\lambda\sigma[\sigma(\psi)\cdot or\cdot\phi];}{\lambda y\lambda x\lambda P[P(x)\vee P(y)];}}{\lambda\psi\lambda\sigma[\sigma(\psi)\cdot or\cdot Mary];\ \lambda x\lambda P[P(x)\vee P(m)];\ S\!\upharpoonright\!(S\!\upharpoonright\!NP)\!\upharpoonleft\!NP}}{\lambda\sigma[\sigma(John)\cdot or\cdot Mary];\ \lambda P[P(j)\vee P(m)];\ S\!\upharpoonright\!(S\!\upharpoonright\!NP)}}{\cfrac{\lambda\psi[\psi\cdot\phi\cdot sleep](John)\cdot or\cdot Mary;\ \lambda x[R(sleep)(x)](j)\vee\lambda x[R(sleep)(x)](m);\ S}{John\cdot\phi\cdot sleep\cdot or\cdot Mary;\ R(sleep)(j)\vee R(sleep)(m);\ S}\ \beta\Rightarrow}}{\lambda\phi[John\cdot\phi\cdot sleep\cdot or\cdot Mary];\ \lambda R[R(sleep)(j)\vee R(sleep)(m)];\ S\!\upharpoonright\!((NP\backslash S)/(NP\backslash S_{bse}))}\ \upharpoonleft I^a$$

Now we simply combine our continuation with the regular VP/VP auxiliary, to get the distributed-scope negation reading.

(39)
$$\cfrac{\cfrac{didn't;}{\lambda Q\lambda x[\neg Q(x)];}\qquad \cfrac{\lambda\phi[John\cdot\phi\cdot sleep\cdot or\cdot Mary];}{\lambda R[R(sleep)(j)\vee R(sleep)(m)];}}{John\cdot didn't\cdot sleep\cdot or\cdot Mary;\ \neg sleep(j)\vee\neg sleep(m)];\ S}$$

7.3   Strip-Gapping

One prediction of this analysis is that there is nothing stopping one from Gapping with a Stripping sentence functioning as one of the conjuncts, and indeed this kind of sentence does appear to exist, as in (40) below. *Robin drinks vodka, and scotch too* can be treated as a straightforward Strip, the likes of which we have already seen before. *Robin drinks vodka, and Leslie, gin* also looks like a normal Gapping sentence. We know how to handle those too. The trick is to derive the Stripping sentence such that it ends up as something of type $S\!\upharpoonright\!((NP\backslash S)/NP)$, or in other words, something that would be a sentence if its gap for a medial transitive verb were filled.

The way to do this is relatively simple. Just Strip as normal, but with the exception of using a dummy hypothetical transitive verb instead of "drinks" for the time being. Lambda abstract out the dummy verb at the end of the Stripping operation to obtain the requisite expression of type $S\!\upharpoonright\!((NP\backslash S)/NP)$. Secondly, derive *Leslie...gin* as usual in the Gapping analysis put forth in Kubota and Levine 2016, to also obtain an expression of type $S\!\upharpoonright\!((NP\backslash S)/NP)$. Then, those two expressions are combined using the Gapping operator from Kubota and Levine 2016. Finally, feed in the transitive verb "drink", resolve the generalized conjunction, and β-reduce.

(40)   Robin drinks vodka, and scotch too, and Leslie, gin.

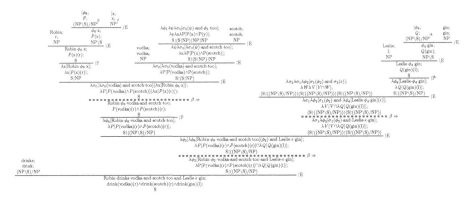

8   Conclusion and Discussion

The analysis presented here extends the analysis of Kubota and Levine's analysis of Gapping in [17] to cover many common forms of stripping, another kind of ellipsis, and shows backwards-compatibility by also obtaining proofs for combined strip-gap sentences. Interestingly, this approach also makes obvious the fact that stripping parallels gapping with respect to the availability of anomalous scope, where certain semantic operators can scope out of their conjunct to take scope over the entire conjunction or disjunction. While this paper cannot give a full account of this phenomenon, evidence from Stripping gives us a larger empirical space in which investigate this fascinating conundrum. In closing, however, there are several minor observations to make about the problem of anomalous scope at this point.

First, as mentioned before, this phenomenon is not restricted to stripping. Example (41-a) demonstrates that similar cases are found in Gapping sentences, a discussion of which can be found in Kubota and Levine [2016]. Example (41-b) is a non-ellipsis example of a raising-to-subject predicate, which also exhibits this behavior.

(41)   a. John can't eat pizza, or Charlie fish.
         (i) $\neg\Diamond(\text{eat}(\text{pizza})(\text{john})\vee(\text{eat}(\text{fish})(\text{charlie})))$
        b. A donkey probably brayed.
         (i) $\text{probably}(\text{brayed}(\text{donkey}))$

Similarly, the fact that most examples in this paper have dealt with negation scoping wide over disjunction should not be taken to mean that is the sole case of the phenomenon. In contrast, (42-a) and (42-b) demonstrate that aside from negation, other modal auxiliaries also demonstrate the same anomalous scoping behavior. Similarly, (42-c) shows that negation can scope wide over conjunction as well as disjunction.

(42)   a. Kim should mow the lawn, or Sandy. - Wide scope necessity
       b. Kim probably mowed the lawn, or Sandy. - Wide scope possibility
       c. The Republican can't win the special election and the Democrat too.
          - Wide scope possibility over conjunction[4]

In addition, there is still plenty of work to be done in relation to the Stripping as well, such as generalizing the Stripping operator, which will involve a formal analysis of the set of possible main functors, a set that only partially overlaps with the possible functors in Gapping. Similarly, a comprehensive analysis of stripping would have to account for the existance of intersentential Stripping, Sprouting, and stripping of embedded subjects, as in (43-a), (43-b), and (43-c) below, respectively.

(43)   a. A: Who went to the store? B: Not John! - Inter-sentential Stripping
       b. I have traveled extensively in my time, but not to Bali. - Sprouting
       c. John would go to the movies with Linda, but I very much doubt anyone
          else. - Sprouting from Embedded Context

References

1. Bach, E.: Tenses and aspects as functions on verb-phrases. In: Rohrer, C. (ed.) Time, Tense, and Quantifiers, pp. 19–37. Niemeyer, Tuebingen (1980)
2. Bach, E.: Generalized categorial grammars and the English auxiliary. In: Proceedings of the Fourth Groningen Roundtable (1983)
3. Bach, E.: Some generalizations of categorial grammars. In: Landman, F., Veltman, F. (eds.) Varieties of Formal Semantics, pp. 1–23. Foris Publications, Dordrecht (1984)
4. Culicover, P.W., Jackendoff, R.: Simpler Syntax. OUP, Oxford (2005)
5. Hankamer, J., Sag, I.: Deep and surface anaphora. Linguist. Inquiry **7**(3), 391–428 (1976)
6. Jayaseelan, K.A.: Incomplete VP Deletion and Gapping. Linguistic Analysis **20**(1–2), 64–81 (1990)
7. Johnson, K.: Few dogs eat Whiskers or cats Alpo. In: Kusumoto, K., Villalta, E. (eds.) University of Massachusetts Occasional Papers (23), pp. 47–60. GLSA Publications, Amherst (2000)
8. Johnson, K.: In search of the English middle field (2004). http://people.umass.edu/kbj/homepage/Content/middle_field.pdf. University of Massachusetts, Amherst. ms
9. Johnson, K.: Gapping is not (VP-) ellipsis. Linguist. Inquiry **40**(2), 289–328 (2009)
10. Johnson, K.: Gapping (2014). University of Massachusetts, Amherst. ms
11. Kubota, Y.: (In)flexibiity of constituency in Japanese in multimodal categorial grammar with structured phonology. Ph.D. thesis, Department of Linguistics, Ohio State University (2010)
12. Kubota, Y., Levine, R.: Gapping as like-category coordination. In: Béchet, D., Dikovsky, A. (eds.) LACL 2012. LNCS, vol. 7351, pp. 135–150. Springer, Heidelberg (2012). https://doi.org/10.1007/978-3-642-31262-5_9

---

[4] Thanks are due to Carl Pollard for these examples.

13. Kubota, Y., Levine, R.: Determiner gapping as higher-order discontinuous constituency. In: Morrill, G., Nederhof, M.J. (eds.) FG 2013, FG 2012. LNCS, vol. 8036, pp. 225–241. Springer, Heidelberg (2013). https://doi.org/10.1007/978-3-642-39998-5_14

14. Kubota, Y., Levine, R.: Pseudogapping as pseudo-VP ellipsis. In: Asher, N., Soloviev, S. (eds.) LACL 2014. LNCS, vol. 8535, pp. 122–137. Springer, Heidelberg (2014). https://doi.org/10.1007/978-3-662-43742-1_10

15. Kubota, Y., Levine, R.: Hybrid Type-Logical Categorial Grammar (2015). http://ling.auf.net/lingbuzz/002313

16. Kubota, Y., Levine, R. (eds.): Proceedings for ESSLLI 2015 Workshop 'Empirical Advances in Categorial Grammar'. University of Tsukuba and Ohio State University (2015)

17. Kubota, Y., Levine, R.: Gapping as hypothetical reasoning. Nat. Lang. Linguist. Theory **34**(1), 107–156 (2016)

18. Kuno, S.: Subjects, theme and the speaker's empathy: a reexamination of the relativization phenomena. In: Li, C.N. (ed.) Subject and Topic, pp. 419–444. Academic Press, New York (1976)

19. Lambek, J.: The mathematics of sentence structure. Am. Math. Mon. **65**(3), 154–170 (1958)

20. Lin, V.: Determiner sharing. In: Billerey, R., Lillenhaugen, B.D. (eds.) Proceedings of the 19th West Coast Conference in Formal Linguistics, pp. 274–287. Cascadilla Press (2000)

21. McCawley, J.D.: Gapping with shared operators. In: Peterson, D.A. (ed.) Berkeley Linguistics Society, pp. 245–253. University of California, Berkeley (1993)

22. Merchant, J.: Fragments and ellipsis. Linguist. Philos. **27**(6), 661–738 (2004)

23. Merchant, J.: Fragments and ellipsis. Linguist. Philos. **27**(6), 661–738 (2004). http://www.jstor.org.proxy.lib.ohio-state.edu/stable/25001944

24. Morrill, G., Valentín, O., Fadda, M.: The displacement calculus. J. Logic Lang. Inf. **20**(1), 1–48 (2011)

25. Morrill, G., Valentín, O.: A reply to Kubota and Levine on gapping. Nat. Lang. Linguist. Theory **35**(1), 257–270 (2017). https://doi.org/10.1007/s11049-016-9336-x. https://link.springer.com/article/10.1007/s11049-016-9336-x

26. Nerbonne, J.: Phantoms and German fronting: poltergeist constituents? Linguistics **24**(5), 857–870 (1986). https://doi.org/10.1515/ling.1986.24.5.857. https://www.degruyter.com/view/j/ling.1986.24.issue-5/ling.1986.24.5.857/ling.1986.24.5.857.xml

27. Siegel, M.: Gapping and interpretation. Linguist. Inquiry **15**, 523–530 (1984)

28. Troelstra, A.S.: Lectures on Linear Logic. Center for the Study of Language and Information - CSLI Lecture Notes 29, 1 edn. CSLI Publications (1992). http://gen.lib.rus.ec/book/index.php?md5=50D60F178D41FA92F246659BC1CC84F8

29. Wurmbrand, S.: Stripping and topless complements. Linguist. Inquiry **48**(2), 341–366 (2017). https://doi.org/10.1162/LING_a_00245

Lithuanian Phrasal Comparatives
Without Covert Syntactic Structures

Elena Vaikšnoraitė[✉]

The Ohio State University, Columbus, OH, USA
vaiksnoraite.1@osu.edu

Abstract. Phrasal comparatives can be analyzed as either involving covert structures or as being directly licensed depending on whether they exhibit clausal properties. In this paper, I show on the basis of Lithuanian phrasal comparatives that clausal-like effects can be enforced without invoking covert structures. I provide an empirically motivated analysis of Lithuanian phrasal comparatives in Hybrid Type-Logical Categorial Grammar. Under this analysis, the clausal-like properties of Lithuanian phrasal comparatives emerge as simple effects of phrasal comparative operators.

Keywords: Comparatives · Covert structure · Phrasal comparatives HTLCG

1 Introduction

This paper explores whether it is possible to provide an analysis of a syntactic construction if it has propositional semantics and clausal-like properties without invoking covert syntactic structures. The question is addressed on the basis of phrasal comparative constructions in Lithuanian (a Balto-Slavic language). Phrasal comparatives in Lithuanian are signaled by the morpheme of *už* 'than'. In (1), the comparison is drawn between the nominative subject *Jonas* 'John', the associate, and the accusative case-marked *Tomą* 'Tom', the standard of comparison. The complement of *už* 'than' must be a single accusative case-marked nominal phrase.

(1)   Jonas   bėga   greičiau už   Tomą   (*bėga).
         John.nom run.prst.3 faster   than Tom.acc run.prst.3

         'John runs faster than Tom (*runs).'

From a generative perspective, phrasal comparatives can be derived via two routes. Phrasal comparatives can be derived from clausal sources through some reduction operation of the relevant part of the sentence (e.g. Bresnan 1973; Lechner 2001, 2004; Merchant 2009). The underlying sentential structure of the sentence in (1) under such an approach is shown in (2a). Alternatively,

© Springer-Verlag GmbH Germany, part of Springer Nature 2018
A. Foret et al. (Eds.): FG 2018, LNCS 10950, pp. 121–135, 2018.
https://doi.org/10.1007/978-3-662-57784-4_7

direct approaches assume that phrasal comparatives are base-generated PPs (e.g. Hankamer 1973, Heim 1985, Merchant 2012), i.e. no unpronounced syntactic structures are posited as shown in (2b).

(2)   a.   Jonas bėga greičiau už [$_{CP}$ Tomą ~~(bėga)~~.]
       b.   Jonas bėga greičiau už [$_{NP}$ Tomą.]

The goal of this paper is to give an empirically adequate treatment of Lithuanian phrasal comparatives that is explicit about the syntax and semantics of Lithuanian phrasal comparatives, as well as word order. Previously, Lithuanian phrasal comparatives have been argued to be underlyingly clausal (Grinsell 2012) based on claims of apparent island sensitivity. Grinsell suggests that Merchant's 2009 clausal analysis of phrasal comparatives in Modern Greek can be extended to Lithuanian. I argue against an ellipsis-based analysis of Lithuanian phrasal comparatives as the analysis does not capture the relevant empirical generalizations. I offer a direct analysis of Lithuanian phrasal comparatives that I choose to implement in categorial grammar in the form of Hybrid Type-Logical Categorial Grammar (HTLCG, Kubota 2010, 2015; Kubota and Levine 2016), though a direct analysis can also be implemented in the Minimalist Program (Vaikšnoraitė 2017). I argue that it is possible to account for clausal-like properties of phrasal comparatives without appealing to covert structure. The analysis proposed in this paper shows that (i) direct licensing of phrasal comparatives straightforwardly captures all empirical generalizations about Lithuanian phrasal comparatives; (ii) the empirical generalizations and clausal-like properties of phrasal comparatives emerge as simple effects of lexical specifications of comparative operators.

2   The Empirical Domain: Lithuanian Phrasal Comparatives

This section reviews three empirical generalizations about phrasal comparatives in Lithuanian that have been previously discussed in the literature (e.g. Ambrazas 2016; Grinsell 2012) and introduces a novel empirical generalization. Lithuanian has several comparative morphemes that combine with different lexical categories to produce comparative meanings. These comparative morphemes are compatible with phrasal comparatives as the examples in (3) show. In (3a), the suffix -*esn*- '-er' attaches to the adjective *greitas* 'fast'. In (3b), the suffix -*iau* '-er' is attached to the adverb *greitai* 'fast'. Nominal comparatives are formed by adding *daugiau* 'more' in front of the nominal *knygų* 'books' as shown in (3c) (for more information about the morphology of comparatives, see Ambrazas 2016):

(3)   a.   Jonas   greit-**esn**-is   už   Tomą.
             John.nom fast-er-m.sg.nom than Tom.acc

             'John is faster than Tom.' (Adjectival comparative)

    b.   Jonas   bėga   greič-**iau** už   Tomą.
           John.nom run.prs.3 fast-er   than Tom.acc

           'John runs faster than Tom.' (Adverbial comparative)

    c.   Jonas perskaitė **daugiau** knygų   už   Tomą.
           John read.pst.3 more   book.pl.gen than Tom.acc

           'John read more books than Tom.' (Nominal comparative)

Now we review some empirical generalizations about Lithuanian phrasal comparatives that were discussed in Grinsell (2012), and introduce a new empirical generalization. We already observed the first empirical generalization in connection with (3), namely that the complement of *už* 'than' is a single accusative-case marked noun phrase. Furthermore, adjectival phrasal comparatives are ungrammatical with measure phrases even if the case-marking requirement is met as the examples in (4) show.

(4)   a.   * Jonas   aukštesnis   už   du   metrus.
           John.nom tall.m.sg.nom than two.acc meter.pl.acc

           'John is taller than two meters.'

    b.   * Jonas   bėga   greičiau už   devynis kilomterus   per
           John.nom run.prst.3 faster   than nine.acc kilometers.acc per
           valandą.
           hour.acc

           'John runs faster than 9 km/h.'

Furthermore, nominal phrasal comparatives are ungrammatical with 'more NP subjects' as exemplified in (5a). Lithuanian shares this restriction with many other Balto-Slavic languages (e.g. Polish, Serbo-Croatian, and Bulgarian, see Pancheva 2009 for more details).

(5)   a.   * Daugiau vyrų   valgo   obuolius   už   moteris.
           more   men.gen eat.prs.3 apple.pl.acc than women.acc

           'More men eat apples than women.'

    b.   Vyrai   valgo   daugiau obuolių   už   moteris.
           men.nom eat.prs.3 more   apple.pl.gen than women.acc

           'Men eat more apples than women.'

The sentence in (5a) is ungrammatical because the associate, *vyrų* 'men', is the subject of the sentence is preceded by the comparative morpheme *daugiau* 'more'. When the object of the sentence is preceded by the comparative morpheme, the sentence is grammatical as shown in (5b). To express the meaning of (5a), a clausal comparative must be used as shown in (6).

(6)   Daugiau vyrų   valgo   obuolius   negu moterų.
        more   men.gen eat.prs.3 apple.pl.acc than women.gen

        'More men ate apples than women.'

A novel contribution of this paper is a description of a previously unnoticed empirical generalization. Phrasal comparatives in Lithuanian are only acceptable if the associate is the subject (which in Lithuanian is marked by nominative case) as shown in (7).[1] The sentence in (7a) contrasts the nominative case-marked subject of the sentence, *Jonas* 'John' and the standard of comparison, *Tomą* 'Tom'. In (7b), the associate is *spurgų* 'doughnuts', a genitive-case marked object.

(7) a. <u>Jonas</u>   suvalgė daugiau spurgų   už   Tomą.
         John.nom eat.pst.3 more   doughnut.pl.gen than Tom.acc

         'John ate more doughnuts than Tom.'

     b. * Jonas   suvalgė daugiau spurgų   už   sausainius.
         John.nom eat.pst.3 more   doughnut.pl.gen than cookie.pl.acc

         Intended meaning: 'John ate more doughnuts than cookies.'

This empirical generalization is further exemplified in (8). The examples show that the sentence with a ditransitive verb *padovanoti* 'to gift' is licensed in a context in which the associate is the subject of the sentence *Jonas* 'John'. The same sentence is ruled out if the associate is the dative object *Marijai* 'Maria'.

(8) a. Context: It is Maria's birthday. John and Tom were both invited to
         the party. John brought three gifts for Maria, and Tom brought two.
         Jonas   padovanojo daugiau dovanų   Marijai   už
         John.nom gift.pst.3   more   present.pl.gen Maria.dat than
         Tomą.
         Tom.acc

         'John gave more presents to Maria than Tom (did).'

     b. Context: Tom and Maria have a joint birthday party. John brought
         three gifts for Maria, and two for Tom.
         # Jonas   padovanojo daugiau dovanų   Marijai   už
         John.nom gift.pst.3   more   present.pl.gen Maria.dat than
         Tomą.
         Tom.acc

         'John gave more presents to Maria than (he did to) Tom.'

In sum, all Lithuanian phrasal comparatives share the following restrictions: (i) the standard of comparison must be a single accusative case-marked NP,

---

[1] A small set of verbs (e.g. *mylėti* 'to love', *nekęsti* 'to hate', *mėgti* 'to like') do not follow this pattern and allow the object to function as the associate:

(i) Jonas myli Agnę labiau už viską.
    John.nom love.prs.3 Agne.acc more than everything.acc
    'John loves Agnes more than anything else.'

Further research is necessary to determine why and under which conditions an object can serve as the associate with these verbs.

(ii) a measure phrase cannot function as the standard of comparison, (iii) the associate must be the subject of the sentence. Furthermore, nominal comparatives are ungrammatical with 'more' NP subjects. Any adequate analysis of phrasal comparatives needs to account for these empirical generalizations.

3   Grinsell's (2012) Analysis of Lithuanian Phrasal Comparatives

The previous analysis of Lithuanian phrasal comparatives is couched within the Minimalist Program. Grinsell (2012) advocates for a clausal analysis of Lithuanian phrasal comparatives by adopting Merchant's 2009 analysis of Modern Greek phrasal comparatives. Grinsell, following Merchant (2009), provides an analysis roughly along the following lines: the complement of *už* 'than' underlyingly is a full clause. The surface form of a phrasal comparative is obtained via TP-ellipsis as shown in (9, the ellipsis site is indicated by angled brackets).

(9)   Jonas   bėga   greičiau už   Tomą   <bėga>.
      John.nom run.prst.3 faster   than Tom.acc run.prst.3

      'John runs faster than Tom.'

The derivation of a phrasal comparative proceeds as follows: first, standard of comparison moves out its base position in the TP to a clause external position, SpecFP, to escape ellipsis. The standard of comparison then moves again to SpecPP, which leads to phrasal-like effects, e.g. case-marking. The preposition *už* 'than' under this analysis is assumed to be embedded in a pP shell (following Matsubara 2000), and the preposition itself moves from P to p. This analysis is exemplified in (10), whereby the standard of comparison, *Tomas* 'Tom', moves to Spec,PP via SpecFP leaving an unelided trace, t_1, in Spec,FP.

(10)   Jonas bėga greičiau
      $[_{pP}$ už$_2$ $[_{PP}$ $[_{DP1}$ Tomą $[t_2$ $[_{CP}$ $[_{FP}$ t_1 $<[_{TP}$ t_1 bėga $]]]]]]]$

The main reason that Grinsell proposes that phrasal comparatives have a covert syntactic structure is that phrasal comparatives seem to exhibit island sensitivities in Lithuanian. He suggests that the phrasal comparative in (11) is ruled out on the basis of relative clause island violation (the example is adapted from Grinsell 2012: 40).[2] The standard of comparison in (11), *Medvedevą* 'Medvedev',

---

[2] The example in (10) is corrected for some grammatical and lexical errors and is written in standard Lithuanian orthography. The original sentence with the phrasal comparative that appeared in Grinsell (2012: 40) is provided below:

(ii)   * Daugiau žmonių kas gyvena valstijoje, kurią valdo Obama už
      More people who live in.the.state which governs Obama.nom than Medvedeva.
      Medvedev.acc
      'More people live in the state that Obama governs than in the state that Medvedev governs.'

contrasts with an nominal phrase that is internal to a relative clause, *Obama* 'Obama'.

(11)   * Daugiau žmonių   gyvena   valstybėje, kurią   valdo
        more   people.gen live.prs.3 country.loc that.acc govern.prs.3
        Obama   už   Medvedevą.
        Obama.nom than Medvedev.acc

    'More people live in the country that Obama governs than in the country that Medvedev governs.'

Island sensitivities would be unexpected under a direct analysis as under such an analysis there is no covert syntactic structure and consequently no syntactic movement. Grinsell (2012) thus concludes Lithuanian phrasal comparatives must be underlyingly clausal.

Under this view, the ungrammaticality of (11) results from a prohibition against unelided island-violating traces. Essentially, when *Medvedevą* 'Medvedev' moves to SpecPP from SpecFP, it leaves an island-violating trace above the elided TP as shown. The island-violating trace makes the sentence uninterpretable at the PF as is schematically shown in (12):

(12) $[_{pP}$ už$_2$ $[_{PP}$ $[_{DP1}$ Medvedevą $[t_2$ $[_{CP}$ $[_{FP}$ *t$_1$ <$[_{TP}$ gyvena valstybėje, kurią valdo t$_1$]>]]]]]]

However, (11) does not constitute evidence for island effects. Recall from Sect. 2 that Lithuanian phrasal comparatives are ungrammatical with 'more' NP subjects. Given that empirical generalization (which Grinsell was also aware of), (11) is independently predicted to be unacceptable because it has a 'more' NP subject. Thus, (11) is not empirical evidence for island effects in Lithuanian phrasal comparatives.

In (13), I present a phrasal comparative that does not violate any of the empirical generalizations laid out in Sect. 2. The phrasal comparative is presented in two minimally different contexts that illustrate that the subject of the main clause, *Jonas* 'John', can serve as the associate, while the subject of the relative clause, *Agnė* 'Agne', cannot. One could attribute this effect to island sensitivity, since *Agnė* 'Agne' is a nominal phrase that is internal to a relative clause. I will suggest in Sect. 4 that these apparent island effects are an epiphenomenon of lexical specifications of comparative operators.

(13)   a.   Context: Agne baked a dozen cookies. John ate four of the cookies, while Tom ate two.

        Jonas   suvalgė daugiau sausainių,   kuriuos   Agnė
        John.nom eat.pst.3 more   cookie.pl.gen which.acc bake.pst.3
        iškepė   už   Tomą.
        Agne.nom than Tom.acc

        'John ate more cookies that Agne baked than Tom ate.'

    b.   Context: Agne and Tom each baked a dozen of cookies for a party. John ate five cookies baked by Agne, and one cookie baked by Tom.

> \# Jonas   suvalgė daugiau sausainių,   kuriuos   Agnė
> John.nom eat.pst.3 more   cookie.pl.gen which.acc bake.pst.3
> iškepė   už   Tomą.
> Agne.nom than Tom.acc
>
> 'John ate more cookies that Agne baked than the cookies that
> Tom baked.'

Grinsell's analysis correctly predicts that the reading in (13b) would be unavailable, which is explained as a relative island violation, since the movement of *Tomą* 'Tom' would leave an unelided island-violating trace (cf. 12).

While the clausal analysis captures the island effects, it does not deal well with the empirical generalizations outlined in Sect. 2. Grinsell suggests (2012: 39) that the reduced clause analysis sketched out above may account for the 'more' NP restriction as such sentences would be ruled because of an unelided island-violating trace. The claim however is not presented in more explicit detail. The sentence in (14) is a simple phrasal comparative that does not involve extraction from an island and thus would wrongly be predicted to be grammatical by Grinsell. The analysis he advocates for thus offers no explanation for why 'more' NP subjects are ungrammatical in Lithuanian.

(14)   * Daugiau vyrų   atvyko   už   moteris.
        more   man.pl.gen arrive.pst.3 than woman.pl.acc

        'More men arrived than women.'

Furthermore, the analysis cannot capture the empirical generalization that phrasal comparatives are incompatible with measure phrases, a fact that Grinsell acknowledges and leaves for future research. Given that the clausal analysis does not capture any of the empirical generalizations listed in Sect. 2, an empirically motivated and formally explicit analysis of Lithuanian phrasal comparatives is necessary.

4   Phrasal Comparatives and Hybrid TLCG

In this section, I develop an analysis of Lithuanian phrasal comparatives within Hybrid Type-Logical Categorial Grammar (Hybrid TLCG), a framework with a flexible mapping between the syntax, semantics and the surface string. I show that the empirical generalizations about phrasal comparatives in Lithuanian can be straightforwardly captured by an analysis that does not assume covert syntactic structures. Under the analysis in HTLCG, the empirical generalizations emerge as simple effects of lexical specifications.

4.1   Hybrid TLCG

In this subsection, I introduce Hybrid Type-Logical Categorial Grammar (Hybrid TLCG). Due to space limitations I will only introduce the most important tenets and assumptions of Hybrid TLCG; see Kubota (2010, 2015), Kubota and Levine (2015, 2016) for a detailed introduction.

In HTLCG, there are at least three atomic syntactic categories: NP, S, and N. Other syntactic categories are recursively built out of these atomic categories via syntactic connectives. There are two directional connectives in HTLCG (forward slash / and backward slash \). Lexical entries consist of tuples: a prosodic component, a semantic component, and a syntactic component. A sample lexicon of Lithuanian is provided in (15). Since Lithuanian is a highly inflectional language, I assume a restricted set of syntactic features (possibly formally represented in terms of subtypes of the underspecified type NP) that I will mark as subscripts of syntactic categories.

(15)   Jonas; john; NP_{nom}
        Tomą; tom; NP_{acc}
        bėga; run; $S\backslash NP_{nom}$
        mato; see; $(NP_{nom}\backslash S)/NP_{acc}$

The intransitive verb *bėga* 'run' takes a nominative NP to its left and returns a declarative sentence S. The transitive verb *mato* 'see' takes two arguments: an accusative NP to its right, and a nominative NP to its left. The difference between directional slashes hence corresponds to the surface word order of the arguments as shown in (16, where ∘ denotes the concatenation operator mapping a pair of string terms to a string).

(16)   Forward slash elimination   Backward slash elimination

$$\frac{a;\ \mathcal{F};\ A/B \qquad b;\ \mathcal{G};\ B}{a \circ b;\ \mathcal{F}(\mathcal{G});\ A}$$

$$\frac{a;\ \mathcal{F};\ B \qquad b;\ \mathcal{G};\ B\backslash A}{b \circ a;\ \mathcal{F}(\mathcal{G});\ A}$$

A sample proof of a simple transitive sentence is provided in (17). The transitive verb *mato* 'see' takes two NP arguments to derive a sentence. By applying the two rules for directional connectives in (16), we obtain the correct surface word order and semantics.

(17)
$$\frac{jonas;\ john;\ NP_{nom} \qquad \dfrac{tomą;\ tom;\ NP_{acc} \qquad mato;\ see;\ (NP_{nom}\backslash S)/NP_{acc}}{mato \circ tomą;\ see(t);\ NP_{nom}\backslash S}/E}{jonas \circ mato \circ tomą;\ see(t)(j);\ S}\backslash E$$

The key feature of HTLCG is that it exploits directional slashes as well as a non-directional (vertical) slash. The non-directional slash as the name suggests is not sensitive to the order of arguments in the syntactic component. The word order is kept track of in the prosodic component via λ-binding. The proof theory of HTLCG, with this syncretic set of implicational connectives, represents a fusion of the type logics proposed in Lambek 1958 (essentially following the formulation in Morrill 1994) and Oehrle (1994), with the further development of the latter in de Groote (2001) and Muskens (2003). The proof theory of HTLCG appears to correspond closely to the intuitionistic non-commutative linear logic outlined in Polakow and Pfenning (1999), with which it shares the right, left and linear implication (here, vertical slash) connectives. The elimination rule for the vertical slash is presented in (18).

(18)   Vertical slash elimination

$$\frac{\text{a; } \mathcal{F}; \text{B} \qquad \text{b; } \mathcal{G}; \text{B}|\text{A}}{\text{b(a); } \mathcal{F}(\mathcal{G}); \text{B}}$$

A sample proof of the same transitive sentence is provided in (19). The transitive verb *mato* 'see' takes two NP arguments to derive a sentence. The only difference this time is that word order is explicitly kept track of in the prosodic component via λ-binding of variables over strings that are indicated as subscripted ϕs in (19), while the syntactic connective is not sensitive to directionality. A vertical slash can be converted to a forward or a backward slash via slanting (see Kubota and Levine 2015 for more information about slanting),

(19) $$\frac{\dfrac{\text{toma; tom; NP}_{acc} \qquad \lambda\phi_1\lambda\phi_2.\phi_2\circ\text{mato}\circ\phi_1; \text{ see; } (\text{S}|\text{NP}_{nom})|\text{NP}_{acc}}{\dfrac{\lambda\phi_2.\phi_2\circ\text{mato}\circ\text{toma; see(t); S}|\text{NP}_{nom}}{\text{jonas}\circ\text{mato}\circ\text{toma; see(t)(j); S}}|\text{E}} \text{ jonas; john; NP}_{nom}}{}|\text{E}$$

In the analysis proposed in the next subsection, I will employ the directional implication rules in (16) and the non-directional implication in (18).

4.2   Analyzing Lithuanian Phrasal Comparatives in HTLCG

In generative analyses of comparatives, adjectival, adverbial, and nominal comparatives are given a uniform analysis, i.e. essentially they have the same derivation. This means that in the nominal comparatives the cardinality of a set of individuals has to be assimilated to the supremum of a set of degrees (see e.g. Bresnan 1973). This is achieved by positing a phonologically null operator MANY which is a function that binds a degree argument to the cardinality of individuals (Hackl 2000). Here I propose that the three types of comparatives are derived through district comparative operators instead, while this results in an expanded lexicon, we do not have to posit phonologically null operator. I will now introduce each comparative operator in turn.

Adjectival Comparative. I propose that predicative adjectival comparatives are constructed with a three place operator in (20). The operator combines with a gradable predicative adjective $((\text{NP}_{nom}\backslash\text{S})|\text{Deg}))$, and two noun phrases. In the prosodic component, subscripted ϕs are variables over strings, while σ is a variable over string to string function. The first argument of the gradable adjective is a degree (Deg), which is phonologically null (ϵ) as per standard assumptions (see e.g. von Stechow 1984). In the semantic component, inequality relation is established between two gradable predicates, where MAX is a function that returns a maximum degree to which a property holds. The semantic component thus can be paraphrased as 'the maximum degree to which x is P exceeds the maximum degree to which y is P.'

(20)   Adjectival comparative operator:
$\lambda\sigma\lambda\phi_1\lambda\phi_2.\phi_1 \circ \sigma(\epsilon)\circ\text{už}\circ\phi_2;$

$$\lambda P\lambda x\lambda y.\textsc{max}(\lambda d.P(d)(x)) > \textsc{max}(\lambda d'.P(d')(x));$$
$$S|NP_{acc}|NP_{nom}|((NP_{nom}\backslash S)|\text{Deg})$$

A sample derivation of an adjectival comparative is provided in (21):

(21) Jonas   aukštesnis už   Tomą.
      John.nom taller   than Tom.acc

'John is taller than Tom.'

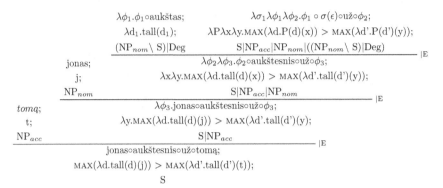

Since the lexical entry in (20) specifies that the operator takes an accusative case-marked NP as its first argument, which is interpreted as the standard of comparison, the first empirical generalization, that the standard of comparison is an accusative-marked NP, has been captured.

The analysis also captures the second empirical generalization. Recall that phrasal comparatives are incompatible with measure phrases like *du metrai* 'two meters' when they are used as the standard of comparison. Semantically, measure phrases have been proposed to be either of type $\langle d \rangle$ in which case they are seen as points on a scale, or as predicates over scale intervals $\langle d, t \rangle$ (see e.g. Schwarzschild 2005 for discussion). Whether Lithuanian measure phrases are $\langle d \rangle$ or $\langle d, t \rangle$ has no bearing on the current analysis, as either way they would be of a wrong semantic type to combine with the comparative operator since it expects an argument of type $\langle e \rangle$. The third empirical generalization that the associate must be the subject is also captured under this analysis, since the gradable predicate that combines with the comparative operator is of syntactic type $((NP_{nom}\backslash S)|\text{Deg}$, i.e. it is a predicate that lacks a subject.

Adverbial Comparative. The adverbial comparative in (22) is a four place operator: it combines with a gradable adverb $((NP\backslash S)\backslash(NP\backslash S))|\text{Deg}$ (for which I will use the shorthand notation $(VP\backslash VP)|\text{Deg}$), a predicate $NP_{nom}\backslash S$ and two noun phrases.

(22)   Adverbial comparative operator
        $\lambda\sigma\lambda\phi_1\lambda\phi_2\lambda\phi_3.\phi_2 \circ \phi_1 \circ \sigma(\epsilon)\circ\text{už}\circ\phi_3;$
        $\lambda f\lambda P\lambda x\lambda y.\textsc{max}(\lambda d.f(P(d))(x)){>}\textsc{max}(\lambda d'.f(P(d'))(y));$
        $S|NP_{acc}|NP_{nom}|(NP_{nom}\backslash S)|((VP\backslash VP)|\text{Deg})$

A sample derivation of an adverbial comparative is provided in (23):

(23) Jonas   bėga   greičiau už   Tomą.
      John.nom run.prs.3 faster   than Tom.acc
      'John runs faster than Tom.'

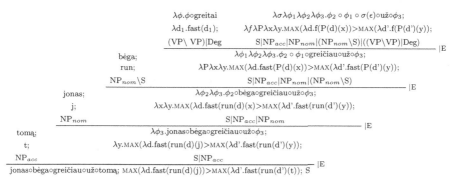

The analysis correctly predicts that the subject of the transitive verb must be the associate, to the exclusion of the object. The syntactic category of the verb must be $NP_{nom} \backslash S$, i.e. a verb that lacks only a single nominative argument to its left. This means that if we have transitive verb like *suvalgė* 'ate' of syntactic type $(NP \backslash S)/NP$, it must first combine with an accusative object. This predicts that (24a) is a well formed sentence with the nominative case-marked subject serving as the associate. Crucially however, it also predicts that the minimally different sentence in (24b) is ruled out as it would predict that *obuolius* 'apples' is interpreted as the subject of *suvalgė* 'ate.'

(24) a. Jonas   suvalgė   sausainius greičiau už   Tomą
          John.nom eat.pst.3 cookie.pl.acc faster   than Tom.acc
          'John ate cookies faster than Tom.'

      b. # Jonas suvalgė   sausainius greičiau už   obuolius.
          John.nom eat.pst.3 cookie.pl.acc faster   than apple.pl.acc

          Intended meaning 'John ate cookies faster than (he did) apples.'
          Predicted meaning 'John ate cookies faster than apples (did).'

Equivalent to the adjectival comparative operator in (20), the lexical entry for the adverbial operator specifies that the accusative case-marked NP argument is interpreted as the standard of comparison. Thus the empirical generalization that the standard of comparison must be a accusative case-marked NP has been captured. The fact that the standard of comparison cannot be a measure phrase is also captured, since it is of a wrong semantic type to combine with the comparative operator.

The Nominal Comparative. The nominal comparative operator in (25) combines with four arguments, two noun phrases, a noun, and an expression of syntactic category $(S \backslash NP)/NP$ (abbreviated as TV for transitive verb). The nominal comparative differs from the other two operators proposed in this section in that the comparison relation is not expressed in terms of degrees, instead the comparison is drawn between cardinality of sets.

(25)   Nominal comparative operator
$\lambda\phi_1\lambda\phi_2\lambda\phi_3\lambda\phi_4.\phi_3 \circ \phi_2$daugiau$\circ\phi_1\circ$už$\circ\phi_4$;
$\lambda P\lambda Q\lambda k\lambda z|\lambda x.P(x)\wedge Q(x)(k)| > |\lambda x.P(x)\wedge Q(x)(z)|$;
$S|NP_{acc}|NP_{nom}|TV|N_{gen}$

A sample proof for a nominal comparative is provided in (26).

(26)   Jonas   suvalgė daugiau saldainių   už   Tomą.
        John.nom eat.pst.3 more   candies.gen than Tom.acc
        'John ate more candies than Tom.'

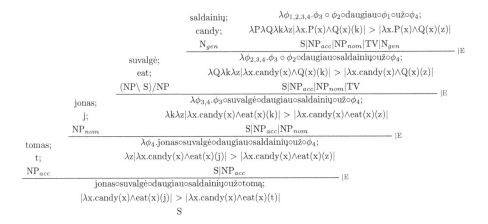

To paraphrase, the meaning of (26) is 'there are more x that are candies that John ate than there x that are candies that Tom ate.' Since the genitive argument of the operator is repeated in the semantic component of both conjuncts, it ensures that the two conjuncts can only differ in their respective subjects. This captures the empirical generalization in Lithuanian that the associate must be the subject. Consequently, sentences like (27) are ruled out on the grounds of being infelicitous. The predicted meaning of (27) would be 'John ate more candies than cookies ate candies', which is infelicitous given that cookies are inanimate.

(27)   # Jonas   suvalgė daugiau saldainių   už   sausainius.
          John.nom eat.pst.3 more   candies.gen than cookies.acc
          'John ate more candies than cookies.'

The proposed operator in (25) also accounts for the empirical generalization that nominal comparatives are unacceptable with 'more' NP subjects. This generalization is captured quite straightforwardly as in such sentences there are simply not enough NPs to saturate all the expected arguments of the nominal operator. Consequently, the sentence in (28) is deemed ungrammatical.

(28)   * Daugiau vyrų   atvyko   už   moteris.
       more   men.pl.gen arrive.pst.3 than women.pl.acc

       'More men arrived than women.'

Furthermore, a welcome consequence of adopting the operator in (25) is that we predict that certain meanings will never arise with phrasal comparatives. For instance, we correctly predict that the sentence in (29, repeated from 13) has only one possible reading, where the associate is *Jonas* 'John'.

(29)   Jonas   suvalgė daugiau sausainių,   kuriuos   Agnė
       John.nom eat.pst.3 more   cookie.pl.gen which.acc bake.pst.3
       iškepė   už   Tomą.
       Agne.nom than Tom.acc

       'John ate more cookies that Agne baked than Tom ate.'
       # 'John ate more cookies that Agne baked than the cookies that Tom baked.'

The semantics of the unavailable reading is presented in (30), this semantics simply cannot be derived given the operator in (25).

(30) $|\lambda x.\text{cookie}(x)\wedge\text{bake}(x)(\text{Agne})\wedge\text{eat}(x)(\text{John})| >$
       $|\lambda x.\text{cookie}(x)\wedge\text{bake}(x)(\text{Tom})\wedge\text{eat}(x)(\text{John})|$

The reason for the unavailability of this reading is that the first argument to nominal comparative operator is a noun, N, the semantics of which is inserted in the semantic component of both conjuncts simultaneously. The consequence of which is that the two conjuncts can only differ in the respective subjects of the verb in the main clause, as shown in the partial proof in (31). Under this analysis, the apparent island effects are just an epiphenomenon of lexical specification of comparative operators.

(31)

$$\frac{\begin{array}{cc}\text{sausainių}\circ\text{kuriuos}\circ\text{Agnė}\circ\text{iškepė}; & \lambda\phi_{1,2,3,4}.\phi_3 \circ \phi_2\circ\text{daugiau}\circ\phi_1\circ\text{už}\circ\phi_4; \\ \lambda x.\text{cookie}(x)\wedge\text{bake}(x)(A); & \lambda P\lambda Q\lambda k\lambda z|\lambda x.P(x)\wedge Q(x)(k)| > |\lambda x.P(x)\wedge Q(x)(z)| \\ N_{gen} & S|NP_{acc}|NP_{nom}|TV|N_{gen}\end{array}}{\begin{array}{c}\lambda\phi_{2,3,4}.\phi_3 \circ \phi_2\circ\text{daugiau}\circ\text{sausainių}\circ\text{kuriuos}\circ\text{Agnė}\circ\text{iškepė}\circ\text{už}\circ\phi_4; \\ \lambda Q\lambda k\lambda z|\lambda x.\text{cookie}(x)\wedge\text{bake}(x)(A)\wedge Q(x)(k)| > |\lambda x.\text{cookie}(x)\wedge\text{bake}(x)(A)\wedge Q(x)(z)| \\ S|NP_{acc}|NP_{nom}|TV\end{array}}|E$$

In sum, this section showed that by virtue of adopting Hybrid TLCG we easily account for all the empirical generalizations of Lithuanian phrasal comparatives, as well as predict apparent island effects.

5 Conclusions

There is a long-standing debate in the literature on phrasal comparatives about whether they are derived from clausal sources or whether they are directly licensed. Most recently it has been suggested that both strategies are instantiated languages with phrasal comparatives (see e.g. Beck et al. 2004; Pancheva 2006, 2009; Bhatt and Takahashi 2007, 2011; Merchant 2009, 2012; Lechner 2015). Different diagnostics have been proposed to help adjudicate between direct and clausal analyses for different languages; essentially if a phrasal comparative exhibits clausal-like properties it should be given a reduced clause analysis. Lithuanian phrasal comparatives have been previously argued to contain covert syntactic structures on account that they exhibit island effects. In this paper, I have shown that while the previous analysis accounts for island effects, it does not capture all empirical generalizations. In this paper, I have developed a direct analysis of Lithuanian phrasal comparatives formalized in HTLCG. Under the analysis proposed here, the three kinds of phrasal comparatives in Lithuanian (i.e. adjectival, adverbial, and nominal comparatives) are derived through distinct comparative operators. The empirical generalizations about Lithuanian phrasal comparatives are captured without the assumption of covert syntactic structures. The analysis indicates that it is possible to account for the apparent clausal properties of phrasal comparatives (i.e. propositional semantics and apparent island sensitivity) without appealing to complex unpronounced structures.

References

Ambrazas, V. (ed.): Lithuanian Grammar. Baltos Lankos, Vilnius (1997)

Bhatt, R., Takahashi, S.: Direct comparisons: resurrecting the direct analysis of phrasal comparatives. Semant. Linguist. Theory **17**, 19–36 (2007)

Bresnan, J.W.: Syntax of the comparative clause construction in English. Linguist. Inquiry **4**(3), 275–343 (1973)

Grinsell, T.: Phrasal and clausal comparatives in Lithuanian In: Proceedings of FASL, vol. 19, pp. 33–49 (2012)

de Groote, P.: Towards abstract categorial grammars. In: Association for Computational Linguistics, 39th Annual Meeting and 10th Conference of the European Chapter, pp. 148–155 (2001)

Hackl, M.: Comparative quantifiers. Ph.D. dissertation, Massachusetts Institute of Technology (2000)

Hankamer, J.: Why there are two than's in English. In: Proceedings of the 9th Annual Meeting of the Chicago Linguistics Society, pp. 179–191 (1973)

Heim, I.: Notes on comparatives and related matters (1985, Unpublished ms)

Kubota, Y.: (In)flexibility of constituency in Japanese in multi-modal categorial grammar with structured phonology. The Ohio State University, Ph.D. dissertation (2010)

Kubota, Y.: Nonconstituent coordination in Japanese as constituent coordination: an analysis in hybrid type-logical categorial grammar. Linguist. Inquiry **46**(1), 1–42 (2015)

Kubota, Y., Levine, R.: Hybrid Type-Logical Categorial Grammar (2015). http://ling.auf.net/lingbuzz/002313. ms

Kubota, Y., Levine, R.: Gapping as hypothetical reasoning. Nat. Lang. Linguist. Theory **34**(1), 107–156 (2016)

Lambek, J.: The mathematics of sentence structure. Am. Math. Monthly **65**, 154–170 (1958)

Lechner, W.: Reduced and phrasal comparatives. Nat. Lang. Linguist. Theory **19**(4), 683–735 (2001)

Lechner, W.: Ellipsis in Comparatives. Walter de Gruyter, Berlin (2004)

Matsubara, F.: p*P phases. Linguist. Anal. **30**, 127–161 (2000)

Merchant, J.: Phrasal and clausal comparatives in Greek and the abstractness of syntax. J. Greek Linguist. **9**(1), 134–164 (2009)

Merchant, J.: Two phrasal comparatives in Greek (2012, Unpublished ms)

Morrill, G.: Type Logical Grammar: Categorial Logic of Signs. Kluwer, Dordrecht (1994)

Muskens, R.: Language, lambdas, and logic. In: Kruijf, G.-J., Oehrle, R.T. (eds.) Resource Sensitivity in Binding and Anaphora. SLAP, vol. 80, pp. 23–54. Springer, Heidelberg (2003). https://doi.org/10.1007/978-94-010-0037-6_2

Oehrle, R.T.: Term-labeled categorial type systems. Linguist. Philos. **17**(6), 633–678 (1994)

Pancheva, R.: More students attended FASL than CONSOLE. In: Proceedings of FASL, vol. 18, pp. 382–399 (2009)

Polakow, J., Pfenning, F.: Natural deduction for intuitionistic non-commutative linear logic. In: Girard, J.-Y. (ed.) TLCA 1999. LNCS, vol. 1581, pp. 295–309. Springer, Heidelberg (1999). https://doi.org/10.1007/3-540-48959-2_21

Schwarzschild, R.: Measure phrases as modifiers of adjectives. Recherches linguistiques de Vincennes **35**, 207–228 (2005)

von Stechow, A.: Comparing semantic theories of comparison. J. Semant. **3**, 1–77 (1984)

Vaikšnoraitė, E.: Phrasal comparatives in Lithuanian (2017, Unpublished ms)

Author Index

Printed in the United States
by Baker & Taylor Publisher Services